KW-222-730

Contents

Introduction

Study the people in the picture. Copy the outline of their shapes onto the first page of your book. If you can't do it free-hand, use tracing paper. All these people are using a lot of energy. Some will be able to go on skating till the ice-rink closes. Others will have to stop to get their breath back. A few will have aching muscles and bruises tomorrow. The skaters are getting hot even though it is cold on the rink. Their hearts are beating faster, their breathing has speeded up, they may be feeling a bit dizzy. Why do our hearts beat more quickly when we skate, or dance, or do any fast exercise? Why do muscles ache and what is a bruise? Why are some people better at balancing than others?

We can't answer these questions from looking at the outlines of the skaters. All we can see is the outside shape and appearance. We can't see what is going on inside, under the skin. We know people have bones and muscles to help them skate. But we can't see the 206 bones, nor can we see the 200 pairs of muscles being used for each movement. And just think how fantastic it is that each skater has 100000 *kilometres* of blood vessels in his body! Work out how many kilometres of blood vessels there are in this picture. How many are there in your family? How many in your classroom?

One man looks as if he may have twisted his ankle. Or he may have torn a tendon. What is a tendon? Where are our tendons? What is their job? Why do they tear? How do they mend?

Most of the time we are fit and healthy so we don't feel curious about what goes on inside us. Then someone we know may get ill, or we may get ill ourselves, and we worry because we don't understand what is wrong. This book will help you to learn about your body and how to keep it healthy. Knowing more about yourself will help you to grow fitter, stronger and more attractive. Good health improves the way we *look* on the outside, our shape and appearance, as well as improving the way we *feel* on the inside.

Biology is the study of living plants and animals. We are human animals. So *human biology* is the study of ourselves. It is finding out all the different things we are made up of and how all these different things work together; keeping us alive, keeping us active, keeping us healthy. 'No man is an island . . .' Each person belongs to a community, a tribe or group of people. We belong to our own particular society. *Health* is not just about ourselves, it is also about the different ways society keeps people healthy, from making sure our tap water is free of germs to providing hospitals to look after us when we are ill. Society is responsible for the health of its people. Each person is just as responsible for his or her own health and for the health of society. Imagine what would happen if one of the skaters had a very infectious disease.

Look back at your copy of the skaters. At the moment you just have some outlines. Don't be tempted to fill in any of the parts of the body you already know. As you go through this course you will fill in the parts *after* you have studied them. You must always check that you know:
a. The *position*, where it is in the body.
b. The *structure*, what it is made up of.
c. The *function*, what it does and how it works.
d. The *hygiene*, how it functions healthily.
e. Any social health problems and what is done to cure them.
(A very useful idea is to find an old length of paper, like wallpaper, lie down on it full length and ask someone to draw around your own body outline. Cut it out and hang it up near your bed or where you do your homework. You can then build up your own personal body chart.)

Human biology and health is a science subject. But it is not a difficult science to learn. People very quickly get interested in knowing more about themselves. One thing though is very important. You will read a great many new words to describe the structure of the body and to explain the functions of each part. **You must learn these new words**. They will be printed in **bold type** so that at the end of each chapter you can re-read them making sure you know their meaning and how to use them. Don't go on to a new chapter till you really know all the words in the last one. (Another very useful idea is to get a notebook especially for these words. This will be your own *glossary*. Write down each word that is in bold type and beside it write out your own meaning of the word, your own explanation.) At the end of the book is an index. Go through it now, lightly marking off the words in bold type which you already know. When you have finished your course, you will be able to put a mark against every word in the index.

It is important you trust your text-book. Read this next part very carefully and remember it. It will save you from getting bothered or worried if you read a fact which doesn't seem to be true, which doesn't seem to fit in with what you have noticed or experienced.

You are an *individual* which means there is no other person quite like you in the world. Because every single person is an individual, we are all different from one another in some ways. Therefore, there are no *absolutes* in human biology and health. For example, if you read, 'The human hand has four fingers and one thumb' you will accept this as true, even though you may know that in rare cases a baby may be born with an extra tiny thumb or finger. (It is removed quite early in the baby's life.) Another example: if you read, 'Smoking is bad for your health' you will accept this as true, even though you may know an old man who has smoked for 30 years and still seems to be healthy. Can you think of an example which you know about?

Science is a living subject which changes as we discover more about ourselves. When science discovers new things, then text-books are changed to keep up with the new information.

What you read in this book will be true about most people, most of the time.

Chapter One

What are we made of?

Protoplasm is the first word you must learn – because protoplasm is the name given to the stuff we are all made of. Protoplasm is life itself; it is living matter. Write the word in your glossary now. If you look it up in your dictionary you will find it means 'the first thing formed'. Protoplasm looks rather like thin grey jelly which hasn't quite 'set'. Most of it, almost four-fifths, is water. The other fifth is made up of many different things.

Protoplasm doesn't flow freely around inside us. We are not great lumps of loose jelly! It is neatly parcelled up into **cells**. A cell is known as a unit of life and we have between 30 million, million and 100 million, million cells in our bodies. Can you write out those numbers in figures? It is hard to imagine such enormous numbers, so we can guess, quite correctly, that each cell must be very small indeed. In fact, a cell is so tiny that it can only be seen by using a **microscope**.

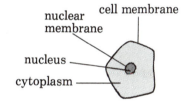

nuclear membrane   cell membrane

nucleus

cytoplasm

Fig 1.1 Diagram of a simple cell

These are the things to look for when you study a cell:
   The **nucleus**, which is the boss or headquarters of the cell. It controls what happens to the cell. The protoplasm in the nucleus is thick because it is closely packed together.
   The **nuclear membrane**, which is the very fine covering around the nucleus.
   The **cytoplasm**, which fills up the rest of the cell. Here the protoplasm is not so thick; it is much looser and more watery.
   The **cell membrane**, which is the very fine outside covering of the cell. It helps to control the things which are allowed to enter and leave the cell.
There are many other things in a cell which are too small to be seen under an ordinary microscope. Inside the nucleus are 23 pairs of, or 46 single, **chromosomes**. The work of the chromosomes is to carry the instructions about the things we have inherited from our parents. The way they do this is explained in Chapter Fifteen. The function of the cell as a whole is to go on doing all the

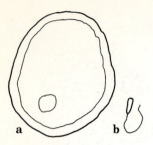

Fig 1.2 a An egg sex cell
b A sperm sex cell

things which keep us alive. This is why a cell is called a unit of life.

We began life as *one* cell. This one cell was made up of a **sperm** sex cell from the father and an **egg** sex cell from the mother.

The sperm and egg mix, or **fuse**, together to make the one cell which was us at the very start of our lives. This fusing together is called **fertilization**. (It is also called 'getting pregnant', 'falling for a baby', and so on.) The fertilized cell then divided into two, these two cells divided into four, four into eight and on and on. How many divisions happened before we were made up of our first million cells? Can you work it out? It is interesting to realize that the number of cells in our body *multiplies* because the cells in our body are *dividing*.

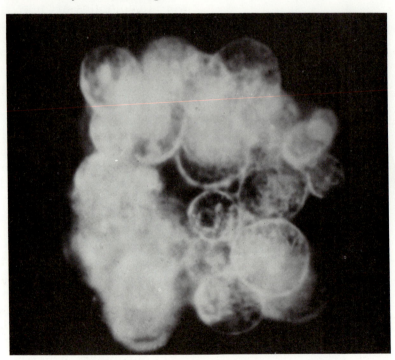

Fig 1.3 A fertilized human egg 3 days old

Figure 1.3 shows us what we looked like three days after our mother was first pregnant – just a bundle of cells rapidly dividing. We didn't look human at all.

Figure 1.4 shows us how different we looked when our mother was 9 weeks pregnant. An unborn baby is called a **foetus**. Study this foetus very carefully. Notice how much growth and development has happened. Compare it with the previous picture. How did the foetus get its shape? Where did the muscles, the blood and the skin come from? Scientists don't know the exact answers but they do know cells carry special information about the work they have

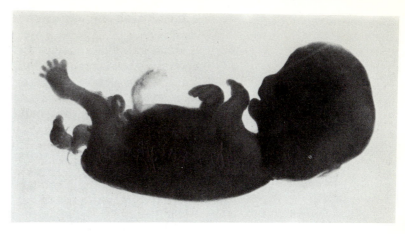

Fig 1.4 A 9 week old human foetus

to do. As cells develop, they change into the shape they need to be to carry out their special function. This is called **cell specialization**.

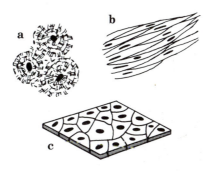

Fig 1.5 a Bone cells
 b Muscle cells
 c Lining cells

Study the cells shown in Fig. 1.5 under a microscope and draw in your book what you actually see.

How cells are grouped together

A group of cells doing a special type of work is called a **tissue**. We have muscle tissue, bone tissue, lining tissue and so on.

Groups of different tissues with a special function are called **organs**. The stomach is an organ; the brain is an organ.

Groups of different organs with a special function are called **systems**. The digestive system deals with food; the muscle system with movement. The systems of the body work closely together though they are studied as separate themes.

Supplies to cells

Each cell must be supplied with the things it needs, e.g. oxygen and foods, in order to carry out all its work. It must also be able to get rid of any waste it makes. Blood brings supplies and takes away waste. But blood doesn't flow in and out of cells. Somehow the cell has to get the things it needs from the blood and pass back its waste into the blood. It also has to be able to pass things around inside

3

the cell itself. It does these things by 'diffusion' and 'osmosis'.

Before a substance can get into a cell, it must dissolve in the liquid in our bodies. This applies to gases and other liquids as well as to solids such as food. Oxygen and carbon dioxide are dissolved in solution long before entering or leaving the cell (page 53). There is plenty of liquid in our bodies to allow solids and gases to be dissolved.

DIFFUSION

Put a lump of sugar into a beaker of water. Don't stir. After a long while the sugar **molecules** (page 86) will have dissolved and spread, or *diffused*, throughout the water. Even though you have not touched the beaker, the sugar molecules have spread till they are evenly distributed in the solution. If you put a drop of ink into some water you can actually watch this happening. Gases are also able to dissolve and diffuse in liquids. The constant movement of molecules to spread evenly through a liquid is called **diffusion**. Liquids are taken to cells and spread, once inside them, by diffusion.

OSMOSIS

In diffusion, dissolved molecules from a strong solution, close to the sugar lump, spread into the weaker solution, the water, to even out the difference in strength. In **osmosis**, the solutions are balanced up by the movement of *water* molecules into the stronger solution.

Osmosis takes place when the weak and strong solutions are separated by a **semi-permeable membrane**. This is a thin film which allows some molecules, like water, to pass through it whilst stopping certain dissolved molecules. Test how osmosis – the movement of water molecules from a weak solution to a stronger one through a semi-permeable membrane – works by doing the experiment shown in Figure 1.6.

The level of liquid will rise up the narrow tube. Why do you think this happens?

Every cell is surrounded by cell membrane which is semi-permeable. If a blood cell, muscle cell or lining cell was put into a strong salt solution, the weaker liquid of the cell would be drawn out by osmotic pressure into the salt solution. The cell would shrink and not be able to work properly. Cells are bathed in **tissue fluid** which helps to keep the correct balance of fluids inside and outside the cell.

The passing of fluids in and out of cells and tissue fluids by diffusion and osmosis is not easy to understand at first. You may need to come back to study diffusion and osmosis later when you have learned more about the workings of the body.

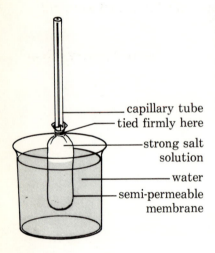

capillary tube
tied firmly here
strong salt solution
water
semi-permeable membrane

Fig 1.6 How osmosis happens

Questions and things to do

Look at as many slides of human cells as you can. Scrape the inside of your mouth to get some lining cells to look at. Study slides of plant cells and write down any differences you notice between these and human cells. Make a list of the animal organs on sale in the butcher's shop. Choose the outline of the smallest skater and fill in, in the correct places, a few of the four cells you have studied.

1. What is protoplasm? Describe it.
2. Draw and label the parts of a simple cell and write down their function.
3. What is meant by cell specialization?
4. Write out these words in order of their size in the body: organ, chromosome, tissue, system, cell.
5. Explain as clearly as you can why a blood cell would shrink and not be able to function properly if it were put in a strong salt solution.
6. What do you think might happen if a blood cell were put in a beaker of distilled water?
7. What is meant by diffusion?
8. Learn all the words and definitions in your glossary. Then test yourself.

Bones and teeth

We think of bones as dead things. Some people don't like the idea of looking at them or touching them. But the bones in our bodies are not dead. *They are alive*. How could we grow taller if our bones were dead? How could a broken arm or leg 'mend' if the bone were dead? Bones are living parts of our bodies.

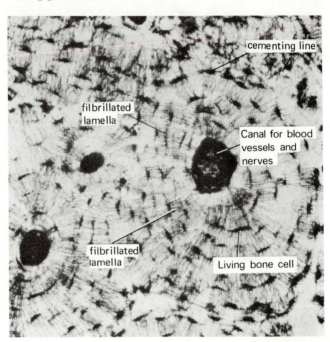

Fig 2.1 Bone cells

Bone tissue

Figure 2.1 shows you what bone tissue looks like. Notice the tiny live cells surrounded by the hard solid stuff which gives bone its strength. The solid part is made of mineral substances: **calcium, phosphate** and **carbonate.** The blood vessels and nerves you can see keep the bone alive and healthy. This bone tissue is from an adult.

Before we were born and during our long childhood, our bones were much softer than they are now. This was because there was a lot of **cartilage** in them. Cartilage is very tough and rubbery; it is the gristle part of the meat

Fig 2.2 X-ray photographs of hands

    a 10 year old child
    b Adult

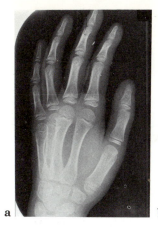 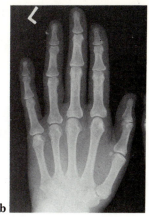

a   b

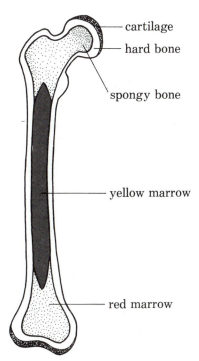

— cartilage
— hard bone
— spongy bone
— yellow marrow
— red marrow

Fig 2.3 A 'typical' bone

A typical bone

we can't chew. But though it is tough, it is not hard and solid like bone. As we grew older the cartilage was changed into bone. This happened quite slowly over the years. Your bones will not be fully hard till you are about 25 years old, and the ends will always be covered with cartilage. The hardening of bones is called **ossification**.

Notice the difference in the X-rays of a young child's hand and an adult's hand shown in Figure 2.2. You can see quite clearly the different amounts of bone. After ossification, bone is made up of 70 per cent hard, non-living matter and 30 per cent living matter. As we get old, our bones may get dry and brittle and lose more of the living matter. Old people must be careful not to fall as their bones snap very easily.

Some cartilage never changes into bone. A few examples are the rings in your neck which hold your breathing tube open, and your nose and your ears. Feel them.

If you look at a skeleton you will see that bones come in all different shapes and sizes. This is because, like cells, bones have special functions and so they develop in a special way for the work they have to do. Before we study the differences, learn what a 'typical' bone is like. One is shown in Figure 2.3. To help you learn this diagram, get a bone from your butcher or from home and find as many of the parts as you can.

Cartilage. This is a smooth covering at the ends of bones to stop them rubbing together.
Outer covering. This brings the nerves and blood vessels to the bone.
Hard bone. This is ossified bone and gives strength.
Spongy bone. Here the bone cells are loosely packed so that the bone will not be too heavy.
Yellow marrow. This has fat cells and is an important store for minerals.
Red marrow. Blood cells are made here.

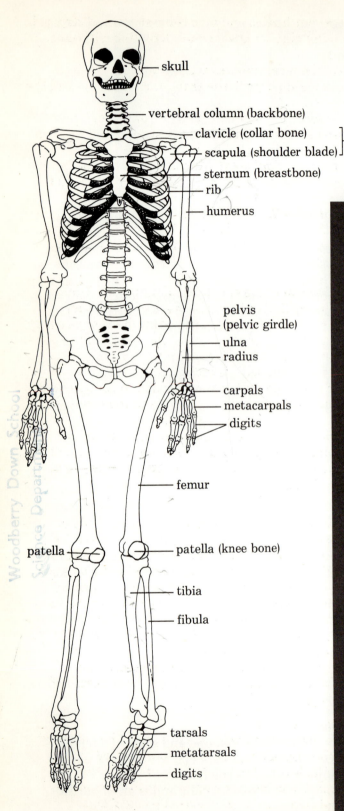

skull

vertebral column (backbone)

clavicle (collar bone)

scapula (shoulder blade) } pectoral girdle

sternum (breastbone)

rib

humerus

pelvis (pelvic girdle)

ulna

radius

carpals

metacarpals

digits

femur

patella

patella (knee bone)

tibia

fibula

tarsals

metatarsals

digits

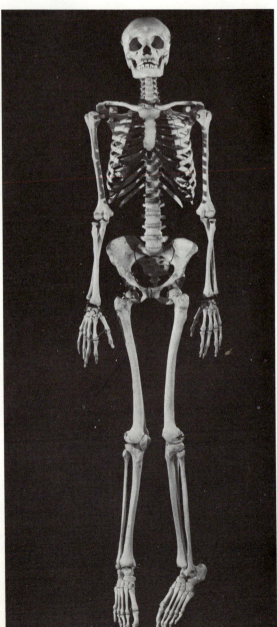

Fig 2.4 The human skeleton

The functions of the skeleton

1. To *support* our bodies and give them shape and strength. (Can you imagine what we would look like without a skeleton!)
2. To *protect* the most important organs of our bodies. (Work out which organs the skull, the backbone and the rib cage protect.)
3. To *allow movement* by being made up of lots of bones which are used as levers. (Could we move if we were one solid bone? Imagine it!)
4. To give *muscle attachment*. Muscles must pull on bone to make us move, so they have to be firmly attached to bone.
5. To *make blood cells* in the bone marrow which are then sent into the blood vessels.
6. To *store* important minerals the body needs, such as calcium salts.

The structure of the skeleton

Study the diagram of the skeleton in Figure 2.4. Then, working in pairs, find out the positions of each other's bones and name each one. (Don't prod too hard or you may be prodded in return.) Make a list of the names of the bones and tick them off as you find them.

You may think you will never be able to learn the structure of the skeleton because it looks so complicated. But it is quite simple once you understand the basic plan.

The basic plan of the skeleton

You could make a copy of the diagrams in Figure 2.5 and build up your own do-it-yourself skeleton. As you learn more about each part of the skeleton keep turning back to this page to check.

THE SKULL

The bones of the skull are flat, with tiny, jagged edges. These edges fit smoothly together to make joints called **sutures**. When a baby is born the skull bones press closer, to make the birth easier for mother and baby. New babies have spaces on their skulls which are not yet joined. These are called **fontanelles**, and there is one at the back and a large one at the top of the head. They close completely by the time the baby is about 18 months old. As the child grows, the skull becomes hard, making a very good protection for the delicate brain inside. Study Figure 2.6. Notice the sutures, the deep eye sockets and the hinged lower jaw.

THE VERTEBRAL COLUMN

This is the name for the backbone or spine. The vertebral column is made up of rings of small bones called **vertebrae**. Each vertebra is piled on top of another to make a column. The biggest vertebrae are at the lower end of the column, and, to give extra strength and support, are fused together. Notice the tiny tail vertebrae.

9

Fig 2.5 a Skull, backbone and ribcage
  b Girdles added to **a**
  c Limbs added to **b**

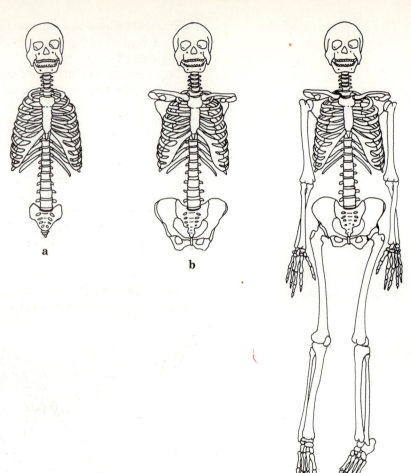

a

b

c

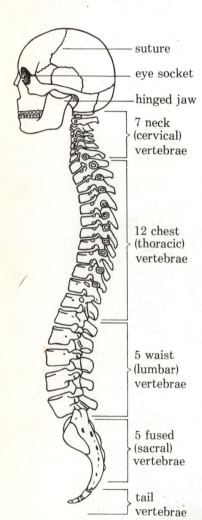

suture

eye socket

hinged jaw

7 neck (cervical) vertebrae

12 chest (thoracic) vertebrae

5 waist (lumbar) vertebrae

5 fused (sacral) vertebrae

tail vertebrae

Fig 2.6 The skull and vertebral column

Fig 2.7 A 'typical' vertebra

They don't seem to be of much use. You can see the vertebrae are different sizes and shapes. Learn the structure of a 'typical' vertebra.

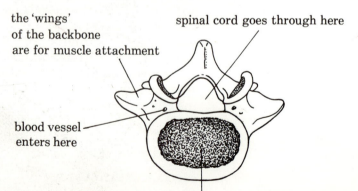

the 'wings' of the backbone are for muscle attachment

spinal cord goes through here

blood vessel enters here

centrum is solid for support and protection

Look at the picture of the nervous system on page 136. You can see the brain continues as a long bundle of nerves right down the back. This is called the **spinal cord**. It is very well protected by the vertebral column.

The vertebrae are separated from one another by discs of cartilage. They act as 'shock-absorbers' or 'cushions'. Have you ever landed with your full weight on your heels? It is a very painful thing to do. Try to work out what you think causes the pain.

THE RIB CAGE

We have 12 pairs of ribs. (One in 20 people have 13 pairs. Mongols sometimes have only 11.) Each pair is joined by cartilage to a vertebra at the back. The breast bone in front is called the **sternum.** The ribs curve round to the sternum. The lower ribs are not joined to the sternum. They are shorter and are joined to each other by cartilage. The eleventh and twelfth pairs of ribs are called floating ribs as they are not joined to anything at the front. Feel all around your rib cage, studying Figure 2.8. carefully. Why do you think it is called the rib *cage*?

The ribs protect two main organs, the heart and the lungs. The ribs are also used in breathing.

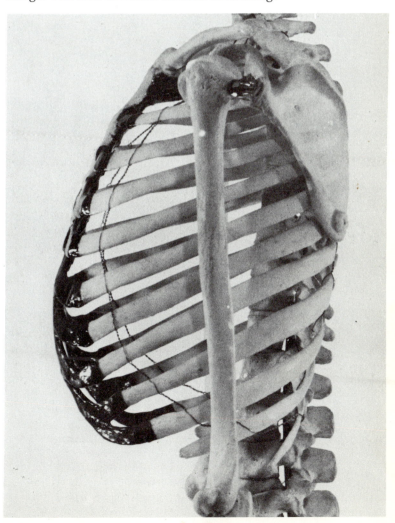

Fig 2.8 The bones of the chest region

THE GIRDLES

The shoulder girdle is called the **pectoral girdle**. There are two pairs of bones, in front the **clavicle**, which is the collar bone, and at the back the **scapula**, which is the shoulder blade. They join the arms up to the rest of the skeleton. They jut out from the rib cage so the arms have plenty of room to swing around.

The hip girdle is called the **pelvic girdle**. It is made up of heavy bones fused together to give strong support to the body and to provide attachment for the huge muscles of the bottom and legs. Notice the sockets for the leg bones and the fused bones of the vertebral column. The pelvic girdle is wider in women than it is in men. The pectoral girdle is wider in men than it is in women. Can you think of any reasons for these differences?

THE LIMBS

Draw the diagrams in Figure 2.9 in your book and learn them.

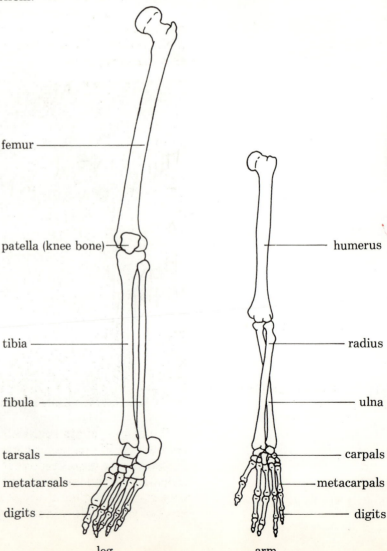

Fig 2.9 The limbs

leg   arm

12

Joints

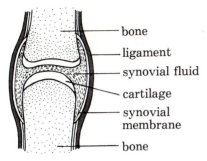

Fig 2.10 A 'typical' synovial joint

A **joint** is the place where two or more bones meet. Apart from the sutures of the brain and the fused bones of the pelvic girdle, most joints have some sort of movement. Study the diagram of a 'typical' moveable joint.

Moveable joints are held together by **ligaments**, which are tough, strong bands of tissue.

The ends of the bones are covered with smooth, slippery **cartilage** which acts as a 'cushion' and a 'shock absorber'.

The ligaments are lined with the **synovial membrane**, which produces a fluid.

The fluid is called **synovial fluid**. It oils the joint, keeping it moist and working smoothly.

So moveable joints are called **synovial joints**.

TYPES OF SYNOVIAL JOINTS

Different joints allow us to make different kinds of movements.

The **ball and socket joints** at the shoulder and hip allows us the most movement. The top of the humerus is rounded and fits neatly into the rounded socket at the end of the scapula. This allows our arm to move in a huge, circling movement.

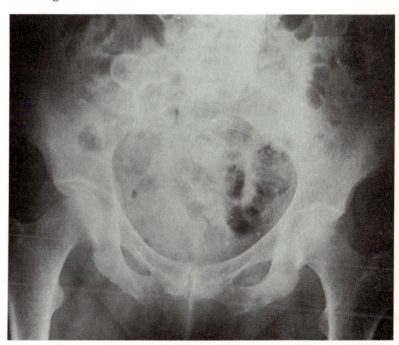

Fig 2.11 X-ray of ball and socket joints at the hip

The **hinge joints** at the elbow and knee allows less movement. We can only move our lower arm in one direction. We cannot move it backwards or sideways, as the bones of the joint act as a brake. The finger joints are very good examples of hinge joints. Try moving the top joint of your little finger sideways or backwards!

The **gliding joints** allow us a twisting movement. Rest your lower arm on the desk, palm upwards. Hold your elbow joint with the other hand. Slowly turn your lower arm so the palm is facing downwards. You can *see* the radius gliding slowly over the ulna.

The **pivot joint** is a special joint between the first two vertebrae of the spinal column. It is the joint which allows us to move our head around on the top of our spine. Move your head and notice how it 'pivots' in various directions. Notice also how the joint acts as a brake to stop your head moving too far sideways or backwards.

The other bones of the vertebral column are not included in the synovial joints. They have only a very slight movement but, as there are so many of them, our backbone is flexible.

LEVERS

The joints of our skeleton not only allow movement, they also act as **levers**. The joint itself is used as the turning point, the **fulcrum**, to 'take the strain' of pulling one bone nearer another, or away from another. The muscle which pulls the bone is attached quite close to the joint. Look at Figure 2.12 and you will see how this also helps the joint to be a very efficient lever.

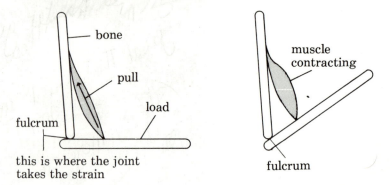

Fig 2.12 Diagram of a joint as a lever

this is where the joint takes the strain

Health and hygiene of the skeleton

Bone is made from calcium and phosphate. But it cannot be made unless we have enough vitamin D in our bodies to help the calcium and phosphate become bone. Look up the foods which are rich in these three things. Our skins can also make vitamin D in sunlight, so people in hot countries can get enough of it. But what about people in cold countries, living in overcrowded cities, where most of the sunlight is blocked out? What should they do to make sure they have enough vitamin D? People in hot countries have dark skins to protect them from the fierce sun. If they move to a cold climate they may not be able to make enough vitamin D. What should they do about their diet to make sure this doesn't happen? Lack of calcium salts and/or vitamin D causes **rickets** in children and bone softening in adults. Vitamin C is also necessary for healthy bones and teeth.

We know that babies and small children have a great deal of cartilage in their bones. This is helpful because they have to learn to crawl, to walk and then to run, and while they are learning they fall down – many, many times. They do not break their bones, as adults would if they kept falling over, because of the amount of cartilage in them. Older children can and do break bones if they fall from a height. But, quite often, they have a **greenstick fracture**. This means the bone bends or cracks a little, rather like a young twig does; it doesn't snap or break as a twig from an older tree would do.

All our growing is done before we are born and during our childhood. This is why it is so important that

Fig 2.13 A child with rickets

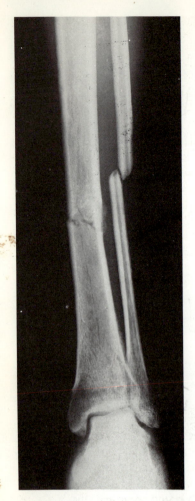

Fig 2.14 Fractures of the lower leg bones

pregnant women, babies, children and teenagers get enough of the right foods to help bones grow straight and firm, and to harden or **ossify** them. Bone growth switches off in the late teens, at age 18 in boys and 16 in girls. These are average ages, so you may go on growing after this age or you may have finished growing already. Bones finally ossify before our twenty-fifth birthday. We get our height and the weight of our bones from our parents (if we have the right diet, of course).

A broken bone is called a **fracture**. This must be 'set' by a doctor. It is then firmly wrapped in plaster so that the two ends of the bone can grow together or 'knit' again. It takes a long while for this to happen but the fact that bones *do* mend will help you to remember they are living things.

A **sprain** is damage done to one of more of the ligaments which hold our bones at the joint. This happens when we twist a joint or fall suddenly onto it. The ligament is torn from the bone and this is very painful. Sprains should always be treated by a doctor. A suspected sprain, or when the ligament is only slightly damaged, can be treated by wrapping the joint in damp cloths to keep down the swelling and by supporting the joint in a sling or, if it is an ankle joint, resting the leg on a stool.

As we grow old, our bones break more easily and take much longer to heal. For this reason, elderly people should never be hurried; they should be allowed to take their time when walking or moving about. Special care should be taken to see there are no slippery floors or mats, especially in the bathroom and kitchen. Do offer to help, if you think he or she might be glad of a strong young person's aid. But try not to embarrass an elderly person by making too much fuss. Be as tactful as you can about helping.

The teeth

Say 'cheese', looking into a mirror. What do you see? Beautiful white teeth and healthy gums? Or dirty, stained, crooked teeth with unhealthy gums? Have you a sweet, fresh mouth? Or a cave of bad breath, smelling of rotting food and decay?

Beautiful teeth are lovely to look at. They need as much care and attention as you give to your hair. Ugly teeth can happen from bad luck (you might have broken a front tooth), or from not caring about them, or from dirty habits.

Before a horse dealer buys a horse, he examines its teeth and gums to see how healthy it is. In quite a different way, we sometimes guess what people are like by how well they look after their teeth. Caring for your teeth is not only necessary for your health, it can also affect the way people think of you. If you don't like your teeth, or if you

Fig 2.15

feel ashamed of them, you can have them fixed quite easily. Go to your dentist now, before you are grown-up and have to pay towards the treatment. You can have them straightened, scraped, filled, cleaned – whatever needs to be done. And then you must care for them yourself. It's up to you whether you can show lovely teeth when you smile.

TWO SETS OF TEETH

We have two sets of teeth during our lives, our **milk teeth** and our **permanent teeth**. Milk teeth begin to develop in the jaws of the unborn baby. They usually come through, or **erupt**, between the ages of 6 and 24 months. There are only 20 of them and they have short roots. They begin to fall out, and the permanent teeth come through, from about 6 years old. By the age of 18, all 32 permanent teeth are usually through. Have you got your 'wisdom' teeth, the heavy ones at the back, yet?

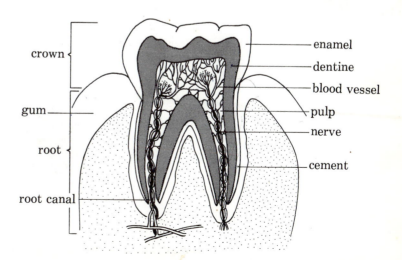

Fig 2.16 A 'typical' tooth

A 'TYPICAL' TOOTH

Like bones, our teeth are different shapes and sizes according to the work they have to do. Draw Figure 2.16 and learn the structure of a 'typical' tooth.

Enamel is a very hard-wearing, non-living covering over the crown of the tooth. It protects the tooth and makes a strong surface for biting.

Dentine is like bone and is made up of calcium salts. It makes the tooth heavy and solid to help grind the food.

Pulp is soft, like its name. It has blood vessels to keep the tooth alive and healthy. It has nerve endings which feel heat and cold. And pain!

Roots hold the teeth firmly in the jaws. The back teeth have two or three roots as they do the heaviest work.

Cement is a thin layer which covers the dentine of the roots. It fixes the roots, as you would expect cement to, firmly to the jawbone.

Tooth specialization

Like cells and bones, teeth develop into different shapes and sizes for the different work they do. Look at your teeth in a mirror then feel the differences between them.

The main functions of teeth are to bite off food and then to chew it, grind it and crush it till it is pulpy and ready to be swallowed. Study the diagram of one half of the jaws, imagining you are eating a stick of celery or raw carrot.

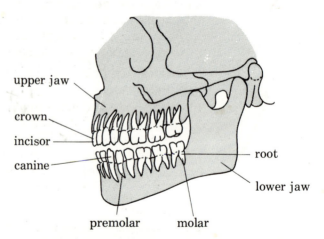

Fig 2.17 Diagram of the four types of teeth

This will help you to learn tooth specialization.

1. The **incisors** cut or bite off the food. On each side there are two on the upper and two on the lower jaw. This is written in a short-hand way like this: $I\frac{2}{2}$.

2. The **canines**, which are slightly pointed like a dog's teeth, grip the food and help tear it off. On each side there is one on the upper and one on the lower jaw, i.e. $C\frac{1}{1}$

3. The **pre-molars** have flattened tops called **cusps**.

18

These help to grind the food. On each side there are two on upper and two on the lower jaw, i.e. $P\frac{2}{2}$.

4. The **molars** are the heavy back teeth which have larger, flattened cusps. They grind, chop and crush the food. On each side there are three on the upper and three on the lower jaw, i.e. $M\frac{3}{3}$.

The **dental formula** is a short way of writing down the number of teeth:

$$I\frac{2}{2} \quad C\frac{1}{1} \quad P\frac{2}{2} \quad M\frac{3}{3}$$

Because it only describes one half of the jaws, we must remember to multiply the formula by two to get the right number of teeth, which is 32.

Health and hygiene of the teeth

In the Western world, only about 2 per cent of people have really healthy teeth with no fillings. This means that 98 per cent of people have filled, decayed or missing teeth. At 40, 50 per cent of people have no teeth left at all! By the time the average child goes to school, he has three or more decayed teeth!

These are dreadful facts. Think of all the pain and misery caused by toothache. Think of all the time and money it costs to keep our dentists so busy and so overworked. Why do we get bad teeth? We know the Eskimos don't even have a word for toothache!

It's believed we get so much tooth trouble because of our diet. The Eskimos eat raw or dried meat and fish, which they have to chew and chew. They eat tough vegetables which are not softened by cooking. We eat soft, sticky mashed-up foods like cottage pie and stewed apples. We also eat lots of sugary things like sweets, cakes, biscuits, ice cream. They don't have trouble with their teeth. We do.

FLUORIDE
Fluoride, added to drinking water, does seem to protect teeth from decay. Would it be sensible to put fluoride into all the drinking water? Some people don't like the idea of having things added to our water. They're afraid if we let one thing be added, all sorts of other things we don't know about might be put in as well. What do you think?

GUM DISEASE
This can be caused by a lack of vitamin C in our diet. We get vitamin C mainly from fresh fruit and vegetables. Lack of vitamin C causes a disease called **scurvy**, making our gums sore and tender and bleeding. There are other diseases of the gum and they must all be treated quickly, as teeth may get loose and fall out.

MILK TEETH
Some parents don't bother about their children's milk teeth as they fall out anyway. This is a serious mistake because if the milk teeth come out too soon, the permanent

teeth may grow down in the wrong place. This means the teeth will be crooked and the child will have to wear a brace to have them straightened.

RULES FOR HEALTHY TEETH

1. A pregnant woman must have a balanced diet so that her unborn baby's teeth develop properly.
2. Parents must train small children to look after their teeth.
3. Visit the dentist every 6 months from an early age.
4. Always brush teeth after meals and last thing at night.
5. At school, clean teeth by eating an apple or, even better, any raw vegetable.
6. Rinse teeth with water, forcing it between the teeth.
7. The toothbrush you use must be soft bristle. Get a new one when the old one looks like this.
8. Use toothpaste with fluoride added. It does help some people.
9. Brush teeth upwards and downwards not sideways, to get at the sticky foods trapped in the cracks.
10. Brush from the gum *downwards for top teeth*, and *upwards for bottom teeth*. If you brush up and down you push back the gum which lets in the germs and, when you are older, makes you look 'long in the tooth'.
11. No chewing gum unless you have perfect teeth. It's so sticky it loosens the fillings and they fall out.
12. Don't use your teeth to open things.
13. Lots of chewing forces the teeth to work, bringing fresh blood to the tooth and keeping the gums healthy.
14. Cut down on eating sweet things.

FALSE TEETH

People who lose their teeth are fitted with **dentures**, false teeth. This is to help them to chew their food, to be able to speak clearly, and to look all right. When we lose our teeth, the gums where the roots have been shrink and the whole of the lower part of the face looks drawn in. This is sad and people who wear false teeth are always sorry they have lost their own. Never lose a tooth if you can possibly get it filled. Have you lost any teeth already? Protect your teeth. Care for them. Get them in shining health and nice to look at now.

Questions and things to do

Examine any bones you can get from your butcher. At home, you could have bones left after a meal of shoulder, leg or neck of lamb, chops or oxtail. Boil them, then study them carefully. Draw them to help you remember what you have learned.

Visit the Natural History part of your local museum.

Fig 2.18 Rules for healthy teeth

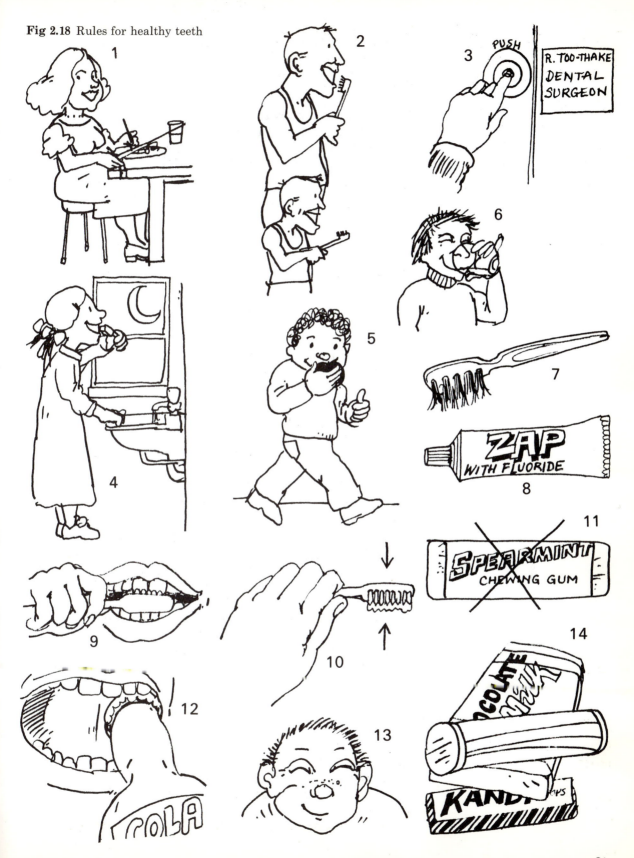

You may see some very interesting skeletons of different animals. Pre-historic skeletons are fascinating to study.

1. How can we tell our bones are not dead things?
2. What is bone made from?
3. What vitamin do we need to help bones develop?
4. What foods are important for bones to grow?
5. Why is vitamin D called the 'sunshine' vitamin?
6. What causes rickets in children?
7. Draw and label a 'typical' bone. Write down the function of each part.
8. Make a list of the functions of the skeleton and learn them.
9. What main organs does the rib cage protect?
10. Why do motor-cyclists have to wear crash helmets?
11. What differences do you notice between the bones of the arms and legs?
12. Draw a 'typical' synovial joint and label it clearly.
13. Learn the four different types of synovial joints.
14. Why can small children bend over so easily?
15. What are the most important times in our lives for bone development?
16. What is a fracture? What is a greenstick fracture?
17. Draw the bones of the leg on one of the skaters.

Look up the dental formulae for sheep, elephant, rabbit and dog. Can you work out why each animal has such different teeth? Think about the food they eat before you answer.

1. If you have already lost any teeth, write out your dental formula.
2. Draw and label a 'typical' tooth. Write down the function of each part.
3. Do you think fluoride should be added to our drinking water?
4. Write about the structure and function of our four different types of teeth.
5. Write down as many things as you can think of which cause tooth decay.
6. Learn all the rules for healthy teeth and gums.

Chapter Three

Muscles and movement

If we could take off our skins we would look something like Figure 3.1. **Muscle** is what we think of as 'flesh' or 'meat'. A delicious, juicy steak is the muscle of a cow. The meat on a pork chop is muscle from between the ribs of a pig.

In our bodies, there are three kinds of muscle:

Heart muscle, which keeps pumping blood around our bodies.

Smooth muscle, which keeps the organs inside us working. Both these kinds of muscle go on working whether we are awake or asleep. We cannot stop them working. They are not controlled by our 'will'. That is why they are **involuntary muscles**.

Voluntary muscles, which are the muscles you see in Figure 3.1, help us to move. They are called voluntary muscles because they are controlled by our 'will' and we can choose whether to move or not. We can choose whether to pick up a pen, turn to look at other students, smile, frown or talk. Voluntary muscles are attached to the skeleton so they are sometimes called **skeletal muscles**.

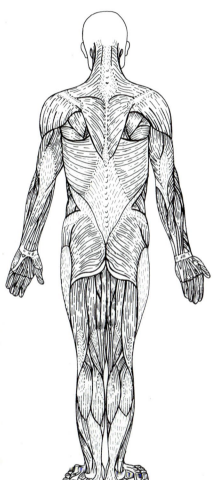

Fig 3.1 The muscles of the body

Fig 3.2 a A few muscle cells (fibres)

b Muscle fibres wrapped together in bundles

The structure of muscle

a

b

Figure 3.2a shows a few muscle cells or fibres. By themselves, they wouldn't be very strong, so lots of muscle fibres are wrapped together in bundles to make a powerful mass of muscle (Figure 3.2b). The muscle on the inside of your upper arm, the **biceps** muscle, is made up of lots of individual muscle fibres wrapped in bundles. The middle part of the biceps bulges out, but at each end it narrows and changes into pieces of tough cartilage. These are the **tendons** and they actually grow into the bones. This means that all our voluntary muscles are very firmly attached to the bones of the skeleton. A muscle must be firmly attached to two or more bones because its job is to move these bones in whatever way we want them moved. A footballer with a 'torn tendon' will not be able to play again until the tendon has grown back and firmly attached itself once more to the bone. This usually takes a very long time.

How muscles work

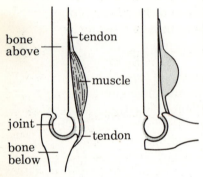

Fig 3.3 How muscle works

Muscles have to pull on bone. They do this by contracting. When a muscle **contracts**, the muscle fibres get short and fat and we can see the bulge. Because the muscle is now small, the bone it is attached to has to be pulled towards it, as shown in Figure 3.3.

Muscles can only pull a bone by contracting. So how does the bone get back to its former position? There must be another muscle to pull it back. And there is. *Muscles work in pairs.* For every one muscle to pull a bone one way, there is a matching muscle to pull the bone the other way. While the first muscle contracts, getting shorter and fatter, the second muscle **relaxes**, getting longer and thinner. And when the second muscle contracts, the first will relax.

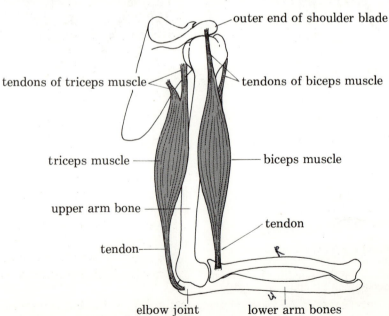

Fig 3.4 Diagram of the biceps and triceps muscles

Study Figure 3.4. The matching muscle for the biceps is the **triceps**. Feel your triceps while your lower arm is resting on the desk. It is longer and thinner than your biceps. Now slowly raise and lower your arm from the elbow, feeling your biceps and triceps in turn. Keep studying Figure 3.4 till you understand exactly how your muscles can make this movement.

When you have done this, move back from your desk and straighten your bent arm down to your side and pull it up again, feeling your biceps and triceps in turn. What difference do you notice in your biceps when your arm is down by your side? Is it fully relaxed or fully contracted?

What you must keep quite clear in your mind is that muscles only work by contracting and pulling on bones. *Muscles only pull on bones.* They cannot and do not push bones.

Nerve supply to muscles

Muscles cannot work until they get messages from the nerves to contract. So muscle cells must have nerve endings. The messages travel from the brain or spinal cord along the nerves to the nerve endings in the muscle fibres.

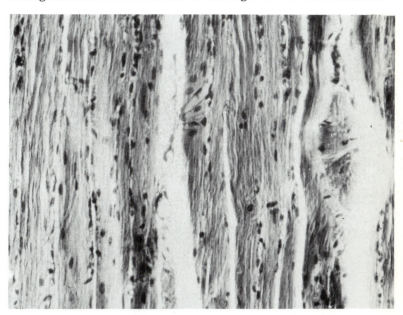

Fig 3.5 Nerve endings in voluntary muscle

They can travel very quickly indeed. Tap your finger as fast as you can on the desk to get some idea of how quickly these messages do travel. If a nerve is damaged no messages can pass to the muscle. The muscle cannot contract until it gets a message to do so. A perfectly healthy muscle will be limp and useless if the nerve carrying messages to it is destroyed. This is called **paralysis** (page 146).

Blood supply to muscles

Muscles need a lot of energy to work. They get this energy from food and oxygen carried in the blood (page 64). While they are working and using up the food and oxygen, they make a lot of heat and they make waste. So muscles must have plenty of blood vessels to bring the energy supplies

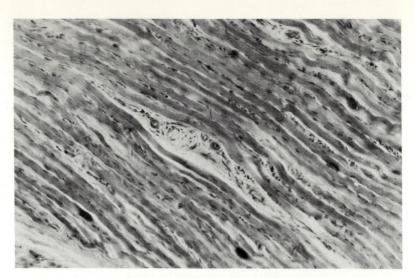

Fig 3.6 Blood vessels in voluntary muscle

and to take away the heat and waste. When you do a lot of exercise, dancing, running, or playing games, you notice your heart is beating faster and your breathing is much quicker. This helps the muscles to get more blood with more oxygen in it. If you over-exercise, if you go on too long without resting, you might get a 'stitch' or cramp. The over-worked muscles are producing a milky fluid, **lactic acid**, which flows over the muscle and causes the pain you feel. As soon as you rest, the blood vessels have more time to take away this waste and so the stitch or cramp goes.

Movement and muscle tone

Picking up a pencil uses at least 12 pairs of muscles. Taking a step uses more than 200 pairs.

A single muscle cell contracts completely or not at all. This is known as 'all or nothing'. When we make a graceful or finely-balanced movement, messages are sent only to *some* nerve endings. This means only *some* muscle cells in a muscle will contract. The grace, style and skill of a ballet dancer, a footballer or a gymnast are the result of just the right amount of messages being sent to the right muscle cells.

To be as good as this we need superb **muscle tone**. To understand what muscle tone is, think of a well-tuned piano. The wires between the keys and the hammers must not be too loose nor too tight; they are in a state of very slight tension. All our voluntary muscles should be in this state of very slight tension, ready and waiting to be used. Even when we are resting, a few of our muscle cells will be slightly contracted. Rest your hand on the desk, palm upwards. Are your fingers stretched out or clenched into a fist? When your hand is at rest, the fingers will be slightly curled inwards. It is muscle tone which keeps your hand in this position. It is muscle tone which holds our body in position when we are not moving. Plenty of exercise helps to give us firm muscle tone. Without exercise our muscles become slack or stiff. When we lose good muscle tone, our movements become clumsy, awkward and without style.

Health and hygiene of the muscles

POSTURE

One hundred years ago, your posture was called your 'carriage', meaning the way you carried yourself. Men were thought handsome if they had a good carriage. Women and girls were called beautiful if they moved well, sat tall and had 'swan-like' necks. As children, they had to practise walking with heavy books on their heads to make sure they would develop into handsome men or beautiful women.

   Nowadays, models and dancers do exercises to improve their posture. And so do most of us, *when we remember*. Are you sitting up straight at this moment? We are often judged by the way we carry ourselves. Because our posture is under the control of our wills, we show how we are feeling by the way we stand, sit or move.

HABIT

Good or poor posture is a habit. It becomes a habit of our

Fig 3.7

27

minds and our bodies. If you only sit or stand well *when you remember*, and if you find you are slouching or slumped down again in a short time, then you have a habit of poor posture. It is very difficult to change a habit. It takes lots and lots of practice and a long time to re-learn the best ways to sit and stand. An example of this is the muscles which form a strong, protective 'corset' for the soft organs of the abdomen, the lower part of our body. If we sit or stand in a slumped position, these strong muscles get used to being stretched, and the organs inside fall forwards out of place. We look as if we have a small 'pot'. As the years pass, the muscles lose their tone, the organs develop out of place and we find we get a sagging belly which we can't pull in. We develop all sorts of digestive troubles and back troubles, and because we look so lumpy, we wear a corset. Corsets are not necessary for healthy men or women; re-training of the muscles will get them back into shape. You can see how important it is to help growing children develop good posture. Once we develop habits of good posture, we stand and sit well without thinking.

Fig 3.8 Damage done by poor posture

DAMAGE DONE BY POOR POSTURE

Any type of poor posture means we lose muscle tone and get flabby.

Head poked forward – headaches, neck and back strain, early double chin.

Stooping shoulders – shoulder and back ache, slight hunched back.

Shoulders hunched forward – this causes very serious damage because the lungs cannot develop or work properly. Without lots of oxygen, we get short of breath, tired, lack energy and we may get serious chest illnesses. In boys, the chest may stay narrow and not develop or broaden. In girls, the breasts soon droop. Breasts don't have any muscle but the muscles of the chest wall help to keep breasts firm.

Backbone slumped – backache, strain put on the discs, soft organs develop out of place, systems not able to work so well.

Weight on one hip only – tired legs and aching back, soft organs develop out of place, trouble with veins in legs.

EXERCISE

Exercise keeps the muscles toned, the tendons supple, the joints working smoothly and the skeleton developing well. People seem either to enjoy exercise – sports, dancing, swimming, ice-skating – or to hate exercise, sprawling or curling up in a soft armchair instead. For those who hate exercise, compare Figure 3.8 with Figure 3.9.

Fig 3.9

You may have no wish to be a gymnast, footballer or dancer. But do you really want to have flabby muscles, stiff tendons and creaky joints? While you are young? Physical Education in school 'educates' or 'teaches' your whole body to work properly. Don't make the mistake of thinking your P.E. teachers are 'health cranks'. It is likely they are more graceful, more skilful, fitter and healthier than you!

Exercise is not only good for our bodies; it is also good for our minds and feelings. It tones them up too. We all need to 'let off steam' at times. Jealous or angry feelings make us tense with misery. Kicking the cat or attacking someone else's property doesn't get rid of bad feelings. It only adds to them. But playing hard for your team, or racing round and round the school yard, or dancing till you are weary, does help to get rid of upset feelings and moods. After we have really exercised, we feel good. We feel stretched, relaxed and have a glowing sense of well-being.

Feet

STRUCTURE

A foot has 26 bones, 117 ligaments, and many tendons and sets of muscles. The toe bones are shorter than finger bones as they are used for balance. Feet are arched to help carry the weight of our bodies, to act as springs and to be good shock absorbers.

Foot A in Figure 3.10 has very high arches on the inside and outside of the foot. Foot B has very low and weak arches. Foot C has the usual arch, high on the inside only. It is perfectly normal to have any of these three types of arches.

FUNCTION

The foot has two main functions. The first is movement; walking, running, dancing. The second is to support our weight when we are standing still. This is more of a strain than moving. Feet need plenty of exercise to work well. If you have to stand still for a time, move your toes about. Soldiers on parade are taught to do this to stop them getting cramp. Small children are not able to stand still for more than a few minutes.

Foot trouble

Small children have perfect feet, with smooth straight toes. Some teenagers have corns and bunions and ugly feet. Many adults and old people suffer from a lot of foot trouble. Why does this happen? Why don't we keep the straight feet we had as children? We are nearly as careless about the health of our feet as we are about our teeth.

SOCKS AND SHOES

A lot of foot trouble starts from the socks and shoes we wear. Look back to page 7 and note again the cartilage and soft bone in a child's hand. This is also true of a child's feet. So if socks or shoes do not fit, you can understand that the soft bones and cartilage will be squashed out of their proper shape. As more bone develops and hardens, it too will grow out of shape. The child's straight feet become **deformed,** which means twisted out of shape.

'Stretch' socks should be really stretchy. Some of then. are not. Shoes should fit snugly round the heel and hold the

A

high arch

B

low arch

C

usual arch

Fig 3.10 Make a print of the arch of your foot

Fig 3.11 Badly deformed feet

instep firmly. Shoes and socks need to be at least a $\frac{1}{2}$ to $\frac{3}{4}$ size larger than the foot to allow the growing toes to lie flat. Some little girls beg their parents to let them wear fashionable shoes. Parents should know that if they give in, they are helping to deform their child's feet. It is not easy to understand why 'fashionable' shoes are nearly always bad for our feet. Can you think of any reasons why? Most people have to decide whether to be comfortable with healthy feet or uncomfortable with ugly feet but smart shoes.

Notice in Figure 3.11 how bunions have developed from the pressure of tight shoes. Instead of lying straight, the toes are twisted and will rub causing pain and damage. Corns, bunions, blisters and callouses are all caused by badly-fitting shoes.

CARE OF THE FEET

1. Children's shoes must support the foot and be properly fitted.
2. Foot measures should always be used in all shoe-shops.
3. Walk around barefoot whenever it is possible.
4. Make sure your feet get enough exercise.
5. If you already have corns or any foot damage, go to a **chiropodist** or your clinic for treatment.
6. If foot trouble is caused by your shoes, you must not wear that pair again.
7. Remember, your bones do not ossify till you are 25, so if the shoes you are wearing now hurt you, they are deforming your feet.
8. Never, never buy shoes that you think you can 'break in'.
9. Because feet sweat, take special care to keep them and all footwear clean and fresh. If you have really smelly feet, wipe them with surgical spirit and air your shoes when you are not wearing them. Wash socks or tights every night.

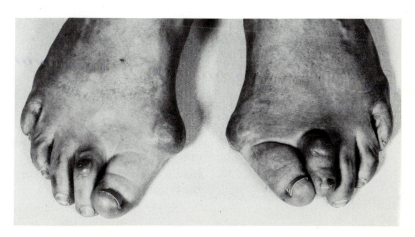

10. Always cut your toenails straight across, as shown in Figure 3.12. Ingrowing toenails are very painful.

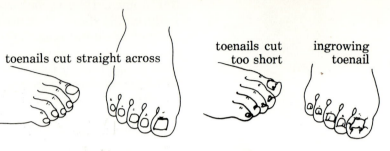

toenails cut straight across   toenails cut too short   ingrowing toenail

Fig 3.12 How to cut your toenails

Questions and things to do

Project. Visit your local health clinic or write to the Society of Chiropodists for more information on foot trouble and how to prevent it. Find out the proper treatment for bunions, corns and callouses. Do a study of shoe-shops and fashion in shoes. Write to a shoe-manufacturer to find out how shoes are made, which are the best, and which are the cheapest materials they use. Why should children's socks be made from wool?

Project. Find out all the sportsfields, parks and playgrounds in your area. Are there enough adventure playgrounds for children? Draw a large map showing all these places. Find out the cost of their upkeep. Make a list of all the sports and activities run by evening classes. Write to the Physical Education Society for information on mountaineering, pony-trekking, canoeing. If you hate exercise, do a study on yoga and one on isometrics.

1. What is meant by 'voluntary' muscle?
2. Why do muscles need a good blood supply?
3. What happens if we over-exercise?
4. What is meant by 'muscle tone'?
5. The Chinese used to bind up their little girls' feet. Find out why.
6. Even if you are a strong swimmer, why should you never go swimming alone?
7. Why are feet arched?
8. Why don't we get corns or bunions on our fingers?
9. Make a collection of pictures of people with good and poor posture. Stick them in your book and write why you think good posture is so important.
10. Test yourself on your understanding of how the lower arm is raised and lowered.
11. Choose a suitable skater and draw in the bone and muscle of the upper arm.
12. Make sure you know the difference between a tendon and a ligament.

Chapter Four

Blood, the heart and circulation

Fig 4.1

Blood

'Blood!' screams a small child, staring in horror at the red, sticky liquid coming from a cut on his hand. Blood is very dramatic stuff. We talk of 'blood and guts', 'our life's blood', 'sweating blood', and so on. Full-grown men have even been known to faint at the sight of it.

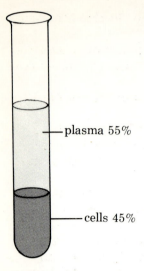

plasma 55%

cells 45%

Fig 4.2

What is it then, this stuff, which travels unseen inside us and makes us panic when it gets out, when we see it?

We know it is red, slightly sticky, and a little heavier than water.
It has a special taste and smell all of its own. A grown man has between five and six litres, a woman about a litre less.
It is a liquid because its main job is to carry things to all the different tissues in our bodies.

Figure 4.2 shows us what happens if we let some blood settle, without clotting.

Blood is made up of 55 per cent liquid, called **plasma** and 45 per cent solids, called **blood cells** or **blood corpuscles**. There are three kinds of blood cells: **red blood cells**, **white blood cells** and **platelets**.

PLASMA

Plasma is a yellow-coloured liquid made up of about 91 per cent water. The rest is made up of things which are being carried to and from the cells and tissues.
Digested food is carried from the intestines (guts) around the body.
Mineral salts are carried around the body.
Urea and other waste is carried to the kidneys.
Carbon dioxide, a waste gas, is carried to the lungs.
Hormones, chemicals which help control the work of our bodies, are carried to where they are needed.
Heat, made in the muscles and liver, is carried all over the body so that we keep an even temperature.
Important proteins, including *fibrinogen*, which helps to clot blood, and *antibodies*, which fight disease, are also in the plasma.
Plasma with the fibrinogen removed from it is called **serum**.

RED BLOOD CELLS
Structure
They have no nucleus.
They are tiny – $5\frac{1}{2}$ million in each cubic millimetre of blood!
They are made in the red bone marrow, mainly of the sternum, ribs and vertebrae.
They only last 3 to 4 months.
They are broken down after this time in the liver and spleen.
Function
The function of red blood cells is to carry oxygen from the lungs to the tissues and cells of the body.
The red colouring of the cell is **haemoglobin**.
Haemoglobin is made of protein and *iron*.
Haemoglobin picks up oxygen in the lungs.
It is then called **oxyhaemoglobin** and looks bright scarlet.

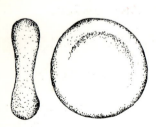

Fig 4.3 Red blood cells

The oxyhaemoglobin is carried in the blood to the tissues and cells needing oxygen.

There it breaks down again into oxygen, which is passed into the tissues, and haemoglobin. The haemoglobin now looks blue.

The red blood cell, with its haemoglobin, travels back to the lungs.

People who live in the highlands and mountains, where there is less oxygen in the air, have more red cells in their blood. This is to make sure they get enough oxygen to live healthy lives.

Anaemia is not having enough red cells, or not having enough red cells *working properly*. People suffering from anaemia cannot get enough oxygen to their tissues so they do not have enough energy. They feel weak and tired, they may look pale, and they may catch other illnesses easily. There are different types of anaemia, but the most usual one is caused from lack of iron. Remember, iron is needed to make haemoglobin. Liver is rich in iron. Look up what other foods would help an anaemic person. A pregnant mother needs extra iron to help her baby's blood. Women who have very heavy periods, who lose a lot of blood at **menstruation**, may need to take iron tablets as well.

WHITE BLOOD CELLS
Structure
They have a nucleus.

They are colourless and can change their shape.

They are larger and less in number than red cells; one white cell for every 600 red cells.

They are not all alike, as they do slightly different work.

Most are made in the red bone marrow; some are made in lymph tissue in the body.

Not enough is yet known about how long they last.

Function
The function of white blood cells is to protect the body from germs.

Dangerous germs, **bacteria** and **viruses** (page 232), get into the body.

The white blood cells travel to where the germs are beginning to cause trouble.

Then they squeeze out of the blood vessels and into the infected part.

They surround the enemy. This is called **engulfing**.

They 'eat up' or digest the enemy. This is called **ingesting**.

Other white cells help to make **antibodies**, which fight germs by making them harmless (page 264). Pus, the whitish stuff we see in a spot or pimple, is made of dead white cells, tissue fluid and germs.

Leukaemia is an extremely serious disease. It is a

Fig 4.4 Two different kinds of white blood cells

chain of bacteria

white blood cell

Fig 4.5 A white blood cell engulfing and ingesting bacteria

cancer of the blood. It is caused when the body makes far too many white cells which are not normal in structure.

BLOOD PLATELETS
These are tiny parts of blood cells which help in the clotting of blood. The function of platelets is to help heal wounds.

SUMMARY OF BLOOD
Blood is made up of 55 per cent plasma and 45 per cent cells. There are red blood cells, white blood cells and platelets. Blood has three main functions:
1. to carry things around the body;
2. to protect the body against disease;
3. to carry heat around the body.
You can understand now that there is nothing 'magic' about blood. But, as the things it carries are essential, it is not surprising we get upset when we lose some of it.

HOW A BLOOD CLOT IS MADE
Our blood must flow freely inside us so blood clots are only made when a blood vessel is cut. **Fibrinogen** and other clotting substances travel harmlessly in the plasma until a blood vessel is damaged. Then a lot of changes happen and **fibrin** is made. Fibrin forms into tiny little criss-cross threads covering the wound. At the same time, blood **platelets** are carried to the wound and get trapped in the network of fibrin. The sticky fibrin with the trapped platelets makes a perfect plug. No more blood can get out. The plug dries and forms a scab. Scabs fall off by themselves, unless they are picked at! Vitamin K and calcium salts are necessary to help the fibrinogen change into fibrin. Make a microscope slide of your own blood and examine the blood clot as it forms.

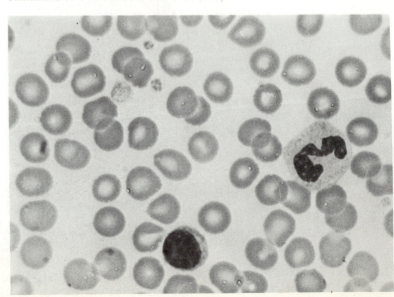

Fig 4.6 The three kinds of blood cells. Can you pick out red and white cells and platelets?

Haemophilia is a rare disease in which the blood clots too slowly or not at all. It is passed down through families, from mothers to sons (page 209).

In older people, a blood clot may form *inside* a damaged blood vessel. A clot in the blood vessels of the heart causes a heart attack. A clot in the brain causes a stroke.

If we cut ourselves deeply, the edges of the wound must be pulled together and stitched or sewn up. This stops us losing too much blood, prevents germs getting into the wound, and makes the area that needs to be plugged with a clot smaller.

BLOOD TRANSFUSION

If we lose too much blood, we need a blood **transfusion** as quickly as possible. Blood is kept ready, stored in refrigerators called **blood banks**. Whole blood can only be stored for about 21 days. Plasma can be dried and stored for longer. A patient may need whole blood or plasma to make up his blood loss. Hospitals need a large store of blood always ready to be used.

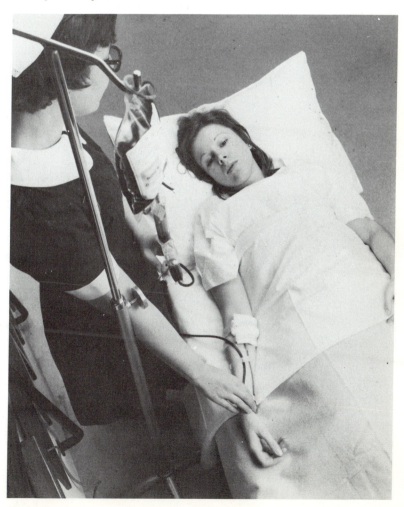

Fig 4.7 A blood transfusion

Most people can be **blood donors**. You must be over 18 and under 65 and not suffering from certain diseases or taking certain medicines. A donor gives about 500 cubic centimetres of blood, then rests quietly and is given tea and biscuits. A visit to a Blood Transfusion Centre only takes about half an hour, and the donor can then go back to work. The blood the donor has given is soon re-made by the body. Will you be a blood donor? Do you think this is an important thing to do?

BLOOD GROUPS
There are many blood groups, but the four main ones are A, B, AB and O. Before we give or are given blood, a blood test is done to see which group we belong to. This is because some blood groups do not mix. If the wrong blood were to be given, the patient would become seriously ill because the blood cells would stick or clump together in the blood vessels.

Group	Can give blood to	Can get blood from
A	A and AB	A and O
B	B and AB	B and O
AB	AB	all groups
O	all groups	O

All blood stored in blood banks is clearly labelled, showing which group it is

The heart

Fig 4.8

A very powerful pump is needed to keep all the important things travelling inside us and to make sure our blood is on the move the whole time. The pump must be able to work non-stop, day and night, year in and year out, from before we are born till the moment we die. The name of this marvellous pump is, of course, the **heart**.

You wouldn't much like a Valentine card with the picture of a real heart on it. It doesn't look very romantic! But we used to believe our feelings came from the heart.

'As he turned away, she felt her heart breaking.' 'His heart was in his mouth with fear.' 'She's got a heart of gold.' But our hearts are made of muscle, not gold. They don't travel anywhere inside us and they don't break! It is true, though, that when we are happy or excited, frightened or angry – when we feel very strongly – we notice our hearts are pounding away, the heart beat getting faster and heavier (page 144). This is probably the reason why we used to believe our feelings came from our hearts. They don't. The heart is made of involuntary muscle called **cardiac muscle**.

POSITION AND STRUCTURE OF THE HEART
The heart lies in the middle of the chest with its lower end

towards the left side. It is almost surrounded by the lungs and is protected by the rib-cage. An adult heart is about the size of a clenched fist and weighs about 400 grams. The tubes you can see are **arteries** and **veins**. Arteries always carry blood *away from* the heart. Veins always carry blood *back to* the heart. On the outside of the heart are small veins and arteries which supply the cardiac muscles. The protective membrane around the heart is the **pericardium**. Work out the meaning of this word.

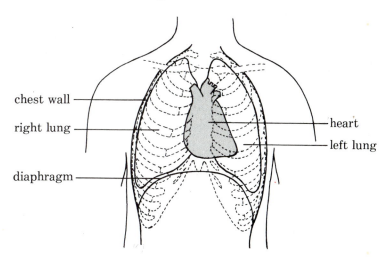

Fig 4.9 The position of the heart

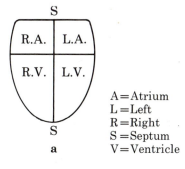

A = Atrium
L = Left
R = Right
S = Septum
V = Ventricle

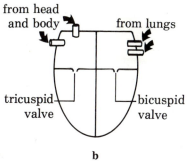

Fig 4.10 a The basic structures of the heart

b Veins and their valves

The basic plan
The **septum** is a wall which divides the heart into two separate pumps. Blood does not, and cannot, pass from the one side to the other. Blood enters and leaves the heart on the same side. The **atria** are rather like two little pockets at the top of the heart. They have thin walls of muscle. The **ventricles** are much larger pockets with thick muscular walls. The left ventricle has much more muscle than the right.

Now find these structures on a dissected heart and learn their names.

Veins and their valves
Veins bring blood back to the heart and pour it into the right and left atria. The **pulmonary veins** come from the lungs, and so the blood is scarlet as it is full of oxygen. Pulmonary veins enter the left atrium. The **inferior venae cavae** are veins which bring blood from the body. The **superior vena cava** brings blood from the head. The blood in these veins looks blue as it has lost most of its oxygen. The venae cavae enter the right atrium. Between the atria and the ventricles are one-way valves. The **bicuspid valve** is on the left and the **tricuspid valve** is on the right.

Find these structures on the heart and learn their names. Look for the tough cords which support the valves.

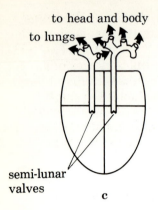

to head and body
to lungs

semi-lunar
valves

c

Fig. 4.10c Arteries and their
valves

Arteries and their valves

Arteries take blood away from the heart. It is pumped out
of the ventricles. The **aorta** leaves the left ventricle,
passes *in front of* the atria and then arches over the top of
the heart. The aorta carries blood rich in oxygen to the
head and the body. The **pulmonary artery** leaves the right
ventricle, passes *over* the atria and then divides to go to the
right and left lungs. The blood in the pulmonary artery
looks blue as it is low in oxygen. At the beginning of both
arteries is another set of one-way valves called the
semi-lunar valves.

Find these structures and learn their names.

HOW BLOOD TRAVELS THROUGH THE HEART

Draw Figure 4.11 in your book and follow the passage of
blood carefully.

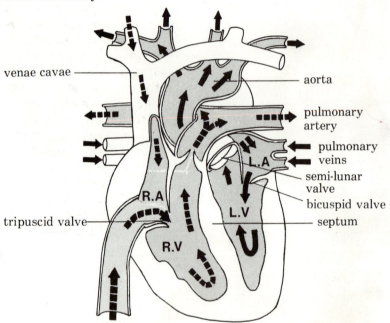

arrows showing flow of blood
■■■■➤ blood with little oxygen
━━━➤ blood full of oxygen

Fig 4.11 How the blood travels
through the heart

Though the heart acts as two separate pumps, both sides
work together. Blood comes flowing back from the veins
and fills up the atria at the same time. The muscles of the
atria walls contract, squeezing the blood into the
ventricles past the open bicuspid and tricuspid valves. The
ventricles fill up with blood.

The strong muscles of the ventricles then contract
powerfully. The bicuspid and tricuspid valves are forced to
slam tightly shut, preventing any blood from returning
into the atria. The semi-lunar valves open and blood is
pumped at high pressure into the arteries. The semi-lunar

valves then slam shut, stopping any blood from trickling back into the ventricles.

(*Remember:* blood which enters the left side of the heart leaves from the left side of the heart and blood which enters the right side leaves from the right side.)

The force of blood entering the aorta is much higher than that of blood entering the pulmonary artery because the left ventricle has far more muscle than the right to pump the blood out harder. The blood in the aorta has to travel to all parts of the body.

THE HEART BEAT

We sometimes think the heart beat is the noise of the muscles contracting. But, in fact, what we hear is the noise of the *valves* being forced to slam shut. Listen to someone's heart beat. You can use a tube of rolled-up paper or you can put your ear against the chest wall. Doctors use a **stethoscope**. You will hear 'lubb' 'dup' then a brief silence. The 'lubb' is the noise of the bicuspid and tricuspid valves closing. The 'dup' is the semi-lunar valves closing. The brief silence is the time lag; the heart is resting for that short moment. If a valve weakens or leaks, doctors can hear the back-wash of blood as it gurgles around. This is one of the causes of 'heart murmur'.

It seems the bigger the animal, the slower the heart beat. A human baby's heart beats about 130 times per minute. At 3 years old, it beats about 100 times a minute, at 12 about 90 times and an adult's heart beats around 70 times a minute. The heart beats much faster when we take exercise, when we have certain illnesses and when our emotions, our feelings, are upset.

Fig 4.12 Heart beat per minute of different sized animals

650   70   25

THE PACEMAKER

This is a small mass of special tissue in the wall of the right atrium. It keeps the heart beating. If the pacemaker stops working properly an artificial one can be used. This is a tiny machine which runs on batteries and it is sewn inside the body. Wires pass electrical currents into the right atrium and keep the heart beating regularly. People who use them are quite proud of their artificial 'tickers'.

BLOOD SUPPLY OF THE HEART

The muscles of the heart have their own blood supply to help them carry on their non-stop work: the **coronary arteries,** which feed the heart. If these arteries are damaged or get blocked by a blood clot, the cardiac muscles cannot get the oxygen and food they need so they cannot go on working. The person collapses and has a heart attack He or she may die unless taken to a doctor or hospital immediately.

BLOOD PRESSURE

Blood is pumped out of the heart into the big arteries under great pressure. This makes sure it will reach our brains, which must have a rich supply of blood. Not only must blood travel upwards, going against gravity, it must also have enough force to be carried to all the different parts of our bodies. As we get older the arteries may harden or get blocked. If you look inside an old waterpipe or kettle you can see the 'furry' deposits. Something rather like this can happen in the arteries. Hardening and/or 'furring' of the arteries causes high blood pressure and can be dangerous.

There are many medical reasons why some older people get high blood pressure or a heart attack and others don't. Older people need to make sure they get enough exercise, stop smoking, cut down on eating too many fatty foods, diet if they are overweight and try to avoid nervous strain. All these things will help to reduce the risk of heart attack and high blood pressure.

THE PULSE

This is not the same as the heart beat, though we take our pulse to find out the rate of the heart beat. Each time blood is pumped out of the heart, a wave of pressure runs along the arteries. It is this wave of pressure we feel, not the blood actually spurting through the arteries. Where an artery is near the surface of the skin and passes over a bone we feel a **pulse**. Find the pulse on your wrist. This may take you quite a bit of time and patience but you must be able to take your own and other people's pulse rate. Other places a pulse can be felt are at the temples, under the neck, the groin and behind the knee. They are called **pressure points**.

BLEEDING

We need not be frightened of blood as children are. After all, blood donors give about two cups of it at a time and it does them no harm. But you can imagine how awful it would look if the same amount was splashed over someone!

When an artery is cut, the blood gushes out in spurts and

is bright red. Arterial bleeding is very serious and must be stopped at once. Don't be frightened. Don't panic. Act quickly and calmly. Send for medical help.

Press down firmly on the wound to slow the loss of blood. Help the patient to sit or lie down and, if the bleeding is coming from an arm or leg, raise that part as high as you can. Keeping the head down makes sure enough blood gets to the brain: raising the wounded part slows the flow of blood to that area. After you have pressed down on the cut for about 15 minutes, put a clean dressing over the wound and bandage it firmly. Keep checking the bandage is not too tight or you will cut off the blood supply to the rest of that area.

How blood travels inside us

This is called the **circulatory system** because blood travels from the heart in two circles. The first circle is a short one. The blood leaves the right side of the heart and goes to the lungs. Here it gets rid of waste carbon dioxide, a little water and heat. It picks up lots of fresh oxygen in the haemoglobin of the red cells. Then it travels back to the left side of the heart. The second circle is much larger.

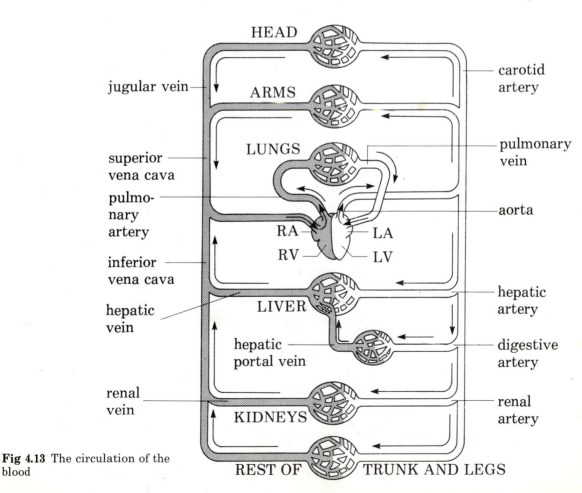

Fig 4.13 The circulation of the blood

Blood, rich in oxygen, goes from the left side of the heart all around the body. As it loses its oxygen it travels back to the right side of the heart.

Blood vessels

Blood doesn't slosh around inside us. It is carried in blood vessels: arteries, veins and capillaries. The aorta branches into many **arteries**. The branches of the arteries get smaller and smaller, rather like the twigs on a tree. **Capillaries** are tiny blood vessels with walls only one cell thick. They join up the smallest arteries to the smallest veins. Small **veins** join up to make larger ones. The large veins join together to form the superior and inferior venae cavae which bring the blood back to the right side of the heart. And so it goes on and on – to the lungs and back, to the body and back – round and round in two 'closed circuits'.

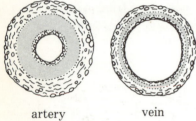

artery   vein

Fig 4.14

DIFFERENCES BETWEEN ARTERIES AND VEINS

Artery	Vein
Takes blood away from the heart	Takes blood back to the heart
Blood travels in spurts	Blood travels smoothly
Has a pulse beat	Has no pulse
Has thick muscle walls	Has thin muscle walls
Has no valves	Many veins have valves
Blood is rich in oxygen (except in pulmonary artery) so looks scarlet.	Blood is low in oxygen (except in pulmonary vein) so looks bluish.

VALVES IN VEINS

Blood flowing back to the heart has lost most of its pressure. To make sure it doesn't slip backwards in the veins there are semi-lunar valves, like those in the heart, to keep it flowing in the right direction. We get **varicose veins** when the valves don't close properly. Varicose veins are unattractive to look at and often quite painful. The best

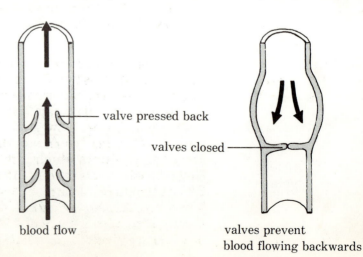

valve pressed back

valves closed

blood flow

valves prevent
blood flowing backwards

Fig 4.15 Valves in veins

way to avoid varicose veins is to take plenty of exercise, as this helps the blood-flow return to the heart.

THE CAPILLARIES

The capillaries form a network of tiny tubes running through the tissues. They are very close to all the cells in the body. They are so narrow that the red blood cells can only pass along them in single file. If you prick your finger, blood *oozes* out. This is because blood travels slowly in the capillaries.

Read again the list of things which the blood carries (page 34). These have to be passed from the blood into the cells. And waste matter from the cells has to be passed back into the blood. Most of this exchange takes place in the capillaries. Figure 4.16 shows how it happens. **Tissue fluid** leaves the blood plasma, passes out through the capillary walls and into the tissues. It carries the things the cell needs. The fluid bathes the cells, passing into them whatever is needed and taking from them whatever has to be collected up. Tissue fluid then returns to the capillaries and forms part of the plasma once more.

capillary wall   white blood cells   tissue cells
red blood cells   tissue spaces

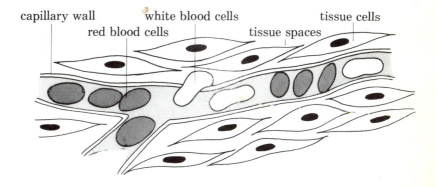

Fig 4.16 How things pass in and out of the capillaries

However, not all tissue fluid passes back into the capillaries. Some is drained into the **lymphatic system**, which will be briefly studied.

The lymphatic system

The tissue fluid which drains into the lymphatic system is called **lymph**. It is like plasma but it has less proteins in it and contains a certain kind of white blood cell. Lymph drains into lymphatic capillaries. These are fine, hair-like vessels which start between the cells and go on to form a network of larger lymph vessels. They join up to make two large **lymph ducts** which take the collected lymph back to the blood vessels entering the heart. One of these ducts carries lymph from the right arm, chest, head and neck. The other, the **thoracic duct**, is very complicated. It carries lymph from the rest of the body back to the heart and it also collects digested fats from the intestines (page 92).

45

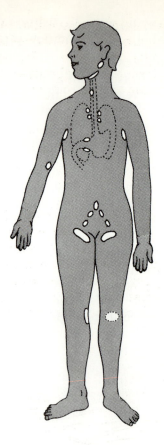

Fig 4.17 The main lymph nodes

The lymphatic system is a 'one-way' system taking fluids from the tissues back to the heart: it is quite different from the 'closed circuit' circulatory system.

Along the lymph vessels are little bumps of lymph tissue called **lymph nodes** or **lymph glands**. These make a certain kind of white blood cell which helps the body to fight disease. They send these white blood cells into the bloodstream. Notice the main glands shown at the armpits, groin and side of the neck. As lymph travels through these glands, harmful things are filtered out. This stops them getting any further into the body and causing more damage. For example, if you have a poisoned finger the lymph glands in your armpit may swell up and feel sore. The **tonsils** and **adenoids** guard the entrance to the lungs and the stomach in much the same way. **Tonsillitis** is an infection of the tonsils caused by bacteria or viruses. The tonsils swell up, are sore and may get covered in white patches. The lymph tissue in the tonsils attacks and destroys the germs before they can get into the body.

The spleen

This is a dark red organ, about the size of a kidney, made up of spongy tissue which has a lot of lymphatic tissue in it. It is to the left of the stomach and is protected by the lower ribs. In some illnesses, such as malaria, the spleen

46

gets very large and swollen. A few of its functions are to help produce white blood cells, to destroy worn out red blood cells and to store the iron from the haemoglobin, which then goes to the liver. However, a person who has to have his spleen removed can live a healthy life without it.

Questions and things to do

1. Prick your finger with a sterilized needle. Put the drop of blood onto a glass slide and examine it under the microscope. Draw what you see. It may take about 20 minutes for fibrin to be formed. Draw what you see (*a*) after 10 minutes and (*b*) after 20 minutes.
2. Study carefully, and then draw, a mammal's heart *before* you cut it open. Label as many parts as you can. When you have opened it, show you know the passage of blood by putting coloured straws down into the veins and up out of the arteries.
3. Visit a Blood Transfusion Centre. Write about everything you notice.

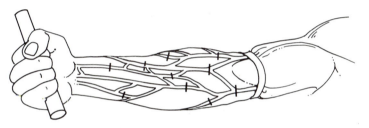

Fig 4.18 Valves in the veins of the arm

4. Copy Figure 4.18. The bandage will stop the blood-flow back to the heart. The veins will stand out. Stroke your forearm down to the wrist. The little bumps are valves. Why do you think they appear?
5. Find out about and make brief notes on the work of William Harvey (1578–1657).
6. Using a stop watch, work out your own pulse rate (*a*) when you are sitting still, (*b*) after 2 minutes' exercise, and (*c*) after 5 minutes' exercise.
7. Choose a skater and sketch in the heart in its correct position.
8. To be of *real* use in any accident or emergency, do a full course of First Aid. Find out what courses there are near you and attend each lecture. They will be of great help to you both practically and in learning more about human biology and health.
9. Find out what is meant by the Rhesus factor in blood.

This chapter has many new words. Learn about the blood first. When you know all the words, *and only then*, go on to the next part. Gradually, you can build up the full vocabulary for the chapter.

1. What is plasma and what is its function?
2. What are red blood cells, where are they made, how long do they last, what is their function, and what happens to them when they are worn out?
3. What is anaemia? What particular foods does a person suffering from anaemia need?
4. White blood cells protect us against disease. Draw one and explain how it works.
5. How is a blood clot made?
6. Learn the structure and function of the heart. (You will have to spend some time on this.)
7. The risk of heart disease can be reduced by . . .
8. Draw and describe the differences between arteries and veins.
9. If a person with B blood group needs a transfusion, which blood groups can he or she be given? Learn the table of blood groups.
10. Write out the pathway of a red blood cell from picking up oxygen in the lungs to delivering it to a cell in the brain. (Work your answer out on rough paper first.)
11. Write out the pathway of some digested food in the plasma leaving the small intestine and being taken to the muscle in your lower leg.
12. Draw and learn the diagram of the circulation of the blood.

Breathing and respiration

If you were asked, 'What do you need to live?' you might answer with a list of things, 'Food, water, air, sleep, exercise' and so on. But if you were asked, 'What do you need to live *at this moment*?' you would say 'Air'. And you would mean the *oxygen* in the air. Oxygen is the most important thing for life. Without it, we die.

'A man can live without food for five weeks, without water for five days, but he can't live for five minutes without air.' (World Health Organization)

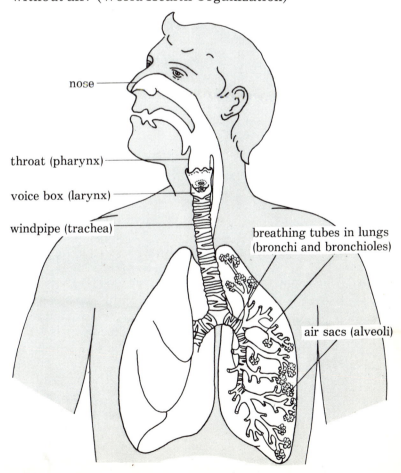

nose

throat (pharynx)

voice box (larynx)

windpipe (trachea)

breathing tubes in lungs (bronchi and bronchioles)

air sacs (alveoli)

Fig 5.1 Structure of the breathing system

The structure of the breathing system

THE NOSE

The **nostrils** are separated by cartilage and soft bone. Inside the nostrils it is warm, damp and slightly hairy. This is because the lining of the nose, like the lining of the whole breathing system, has lots of blood capillaries, a layer of **mucus**, and tiny, hair-like **cilia**. There is also fine hair in the nostrils, which may get thicker as we grow older, to trap the larger bits of dirt and dust we breathe in. Air breathed in through the nose is:

warmed by the heat from the blood in the capillaries;
dampened by passing over the wet mucus;
cleaned by the cilia which sweep dirt onto the mucus;
filtered by the fine hair in the nostrils.

If we are out of breath and panting, or when a cold is at the '*blocked nose*' stage, we breathe through our mouths. Air breathed in through the mouth is cold, dry and likely to be full of dirt and dust. It can damage the soft linings of the throat, breathing tubes and lungs. Our breathing system is very delicate and can easily become infected. Small children need to be gently reminded to close their mouths. Are you a mouth-breather? It may be just a habit left over from your last cold. Try to check that you breathe through your nose.

THE PHARYNX (THROAT)

The pharynx, or throat, is a very busy place. Breathing, keeping air pressure in our ears, and swallowing, all happen in the pharynx. So the traffic between air for breathing, air pressure for hearing and food and water has to be carefully controlled.

a. *Breathing*. Air from the two nose passages goes through the pharynx and down into the beginning of the windpipe, the **trachea**. The trachea is held open by rings of cartilage so that air can pass in and out easily. Raise your chin and feel where the rings begin at the top of your neck.

b. *Air pressure*. There is air in the middle part of each ear to help us to hear. This air comes from the throat and goes to each ear by a little tube. When we travel in an aeroplane, we can adjust the air pressure in our ears by swallowing, which lets some of the air back into the throat, (see page 130).

c. *Swallowing*. Chewed-up foods and liquids are pushed by the mouth and throat muscles through the pharynx and into the beginning of the food tube. But the opening to the food tube is *behind* the trachea, so there has to be a way to stop food dropping into the trachea. Otherwise we would choke. And there has to be a way to stop food and liquid from being pushed up into the back of the nose passages. As we swallow, the

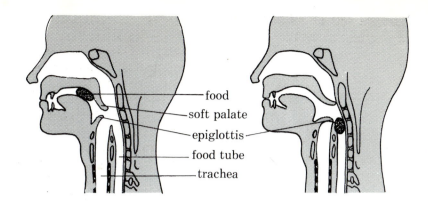

Fig 5.2 Swallowing

food
soft palate
epiglottis
food tube
trachea

soft palate which is at the back of the top of the mouth rises up to block off the opening of the nose. At the same time, the opening of the trachea also moves upwards, which forces a little flap of cartilage, the **epiglottis**, to slide over the entrance to the windpipe, closing it off. So foods and liquids pass safely through the pharynx and down into the food tube. Controlling the traffic of things in the throat happens without our thinking about it.

At some time, though, we've all had something 'go down the wrong way'. This happens because we are talking or laughing or trying to breathe at the same time as we are swallowing, and the signals for controlling the traffic get confused. Nothing must enter the trachea except air, so the coughing and choking and spluttering are necessary to to shoot out the food or water. A sharp pat on the back sometimes helps to get it out.

THE LARYNX (VOICE BOX)

The larynx, or voice box, is the first part of the trachea. It is broader than the rest of the trachea and we can see it bobbing up and down when we swallow. You can find it with your fingers by humming or singing a note and feeling the vibrations. Inside the larynx are tightly stretched cords, the **vocal cords**. When we breathe, the cords are loose and lie back against the walls of the larynx, so air passes in and out quietly. When we speak, the cords come forward again.

We are born able to make sounds, but we have to learn how to speak. We speak as we are breathing out.
Loudness depends on how hard we breathe out.
Pitch depends on how tight or loose the vocal cords are.
Quality depends on the 'amplifiers', which are the mouth, nose, throat, breathing tubes and lung spaces. (When we have a cold or breathing infection, our voices sound dull and flat because the 'amplifiers' are blocked.) Women and children have shorter vocal cords so the sounds they make are higher in pitch. When a young man's voice

'breaks', his vocal cords are getting longer. Because men have longer vocal cords, their voices are deeper.

Laryngitis is an infection of the larynx which causes us to 'lose our voice' for a short while. Laryngitis is known as the singer's curse and the teacher's nightmare!

THE TRACHEA (WINDPIPE)

The trachea, or windpipe, is the round tube, about 10 cm long, taking air from the throat to the beginning of the lung tubes. The larynx is the top part of the trachea. All the breathing tubes are held open by rings of cartilage so they do not collapse. Air can be easily and freely passed up and down.

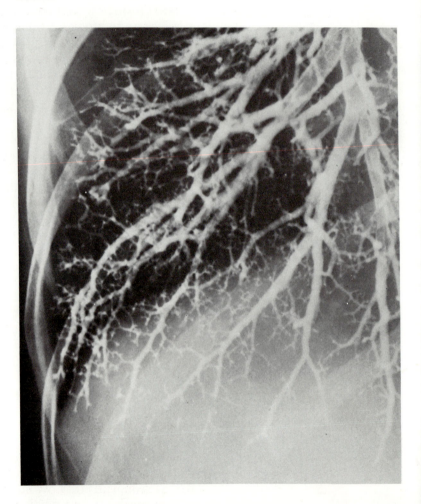

Fig 5.3 X-ray photograph of bronchioles

THE BRONCHI (BREATHING TUBES IN LUNGS)

At the bottom of the neck the trachea divides into two tubes, the **bronchi**. One bronchus goes to the left lung, the other to the right lung. Inside each lung the bronchus divides into smaller and smaller tubes, called **bronchioles**. The tiny bronchioles spread to all parts of the lung. Each one ends in a cluster of little round air sacs. **Bronchitis** is an infection of these tubes.

THE ALVEOLI (AIR SACS)

There are about 300 million tiny air sacs, or **alveoli**, surrounding the ends of the bronchioles. As you can see from Figure 5.4 each alveolus is covered with blood capillaries, like a net. Each alveolus also has a thick lining of moisture. As air comes into the alveolus the oxygen is dissolved, made liquid, in the moisture. Dissolved oxygen can then pass through the wall of the alveolus and into the blood capillaries (page 34). It mixes with the haemoglobin in the red blood cell to make oxyhaemoglobin and is taken back to the heart. Waste carbon dioxide and moisture are passed back from the blood capillaries into the alveolus and are breathed out. This exchange or swapping of gases in the alveoli is called **gaseous exchange** though we usually call it breathing.

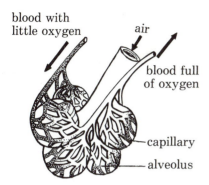

blood with little oxygen

air

blood full of oxygen

capillary

alveolus

Fig 5.4 The alveoli

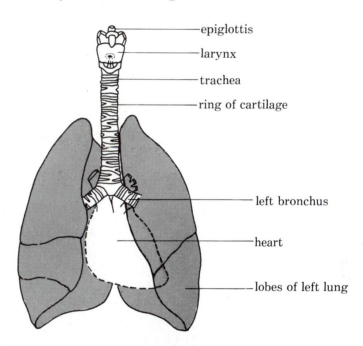

epiglottis

larynx

trachea

ring of cartilage

left bronchus

heart

lobes of left lung

Fig 5.5 The lungs

THE LUNGS

These are two large spongy organs which, with the heart, fill up nearly the whole of the chest. Because we see pictures of lungs from the front, we may think they are just inside the front of the chest wall. It is important to remember our lungs fill the whole of the chest – front, sides and back. The right lung is broader and shorter than the left. It has three lobes and the left has two larger ones. Think about the position of the heart and work out why the lungs are slightly different in shape. Lungs are made up of the end of the bronchi, the bronchioles, the alveoli, the blood capillaries and lung tissue which can stretch like elastic. If all the air sacs in the two lungs were spread out, they would make a breathing surface almost as large as a tennis court! This makes sure there is plenty of space to get all the air we need. **Pneumonia** is a serious disease of one or both lungs.

THE PLEURAE (COVERINGS OF LUNGS)
The lungs are wrapped in two layers of membranes, the **pleurae**. Between the pleural walls is a fluid which acts as a lubricant or oil. **Pleurisy** is an infection of the membranes and fluid. The bones of the rib cage protect the lungs and the heart.

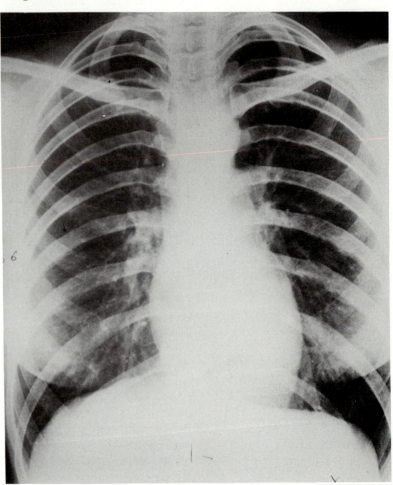

Fig 5.6 X-ray of the chest

Protection against disease

As oxygen is so vital to us, and the organs of the **respiratory** (or breathing) **system** are so delicate, we have special protections against infections which may be breathed in from dirty air.

All the breathing passages are lined with membranes rich in blood capillaries. This means that the air is warmed before it enters the air sacs.

These membranes produce **mucus**, which is thin, slippery, slightly sticky and wet. Mucus moistens the dry air we breathe in. It also traps dust, soot, pollen, bacteria and many other things which might harm us.

The lining of the mucus membrane is covered with masses of tiny little fronds or hairs, called **cilia**. They can only be seen under a microscope. They move to a beat, in an upward direction, rippling like the legs of a millipede, wafting and waving the mucus and dirt up to the throat and nose. This movement goes on the whole time so there is a constant travelling of mucus upwards. We don't feel this happening. We get rid of mucus from our nose by blowing it into a handkerchief. Mucus in the throat is swallowed and destroyed in the stomach. The movement of the cilia is an excellent 'sweeping out' system, keeping our lungs healthy and clean.

Smoking damages the work of the cilia. Smokers and people with colds and chest infections get a lot of mucus, which becomes thick and sticky. It may be called **phlegm**, **catarrh** or **sputum**. Phlegm should be spat out of the mouth into a paper handkerchief and then destroyed. In Britain, phlegm used to be spat out into the streets. There were special bowls called 'spitoons' in the corners of rooms for people to use. We know now the germs in phlegm dry up and get blown into the air we all breathe, so coughing up phlegm in the streets is not allowed though it is still the custom in some other countries.

The common cold

It is probably called the 'common' cold because anyone can get it. Colds are caused by **viruses**, which are the smallest of all germs. They float around in the air and we breathe them in. Within a few days we are coughing and sneezing, we have a runny or blocked-up nose, and maybe a dry, sore throat and a headache. Colds are not serious, except for people with chest trouble, babies and old people. But they are a nuisance. There is no need to stay at home with a cold, but when you go out you must be *very, very* careful not to spread it to other people. Don't be generous! Be mean like Scrooge. Catch every sneeze, every cough, every sniffle in your handkerchief (page 250).

On average, we get about two colds each year, which means some people get more and some people get less. The only chance you have not to catch a cold is to get yourself healthy and *stay* healthy. Keep warm. Wear enough clothes. Eat well. Get plenty of sleep and exercise. We should take care not to let a cold develop into a more serious illness. Some babies and small children seem to catch one cold after another. They get swollen adenoids and infected tonsils and many other complaints. Parents should make sure a child's bedroom is warm all through

THE QUEEN HAS CANCELLED HER ENGAGEMENTS DUE TO A HEAVY COLD

Fig 5.7

Fig 5.8

the night. It is nonsense to have a bedroom window flung wide open on a bitterly cold night. And it is just as silly to send a child to school in a short skirt or short trousers. Not many schools still insist on this sort of uniform in winter. Adults don't go around with bare legs in winter. But some people have the odd idea that you 'harden' a child or 'make a child tough' by sending him or her out in the freezing cold with bare legs, hands and head. You cannot toughen the delicate linings of the respiratory system. You only damage them by breathing in raw and icy air.

Fig 5.9 Don't be so generous with *your* germs

Fig 5.10

The common cold may last up to 2 weeks. If it goes on after that, go to your doctor. Figure 5.10 shows some of the things you can buy from the chemist if you have a cold. These do not, and cannot, kill off the common cold virus. Remember this, because the advertisements hint that they can cure you. They cannot. What they can do is soothe the sore places, unblock a stuffed-up nose, and ease the cold symptoms.

Apart from a cold developing into a disease, the most serious effect is the time lost from work and from study.

Influenza

Quite often, people think they have 'flu when they only have a nasty cold. The symptoms for 'flu are quite different from a cold. There is a feeling of shivery chilliness, headache, muscle and joint pains, great tiredness and a sudden high rise in temperature. The fever may last for 3 days and as it clears up, the symptoms of a cold might start. 'Flu can lead to serious diseases such as pneumonia. People with 'flu must go to bed and stay there. They need lots of fluid, a light diet and a warm room.

Fig 5.11

There is no need to call the doctor for 'flu, but if you are worried telephone and describe the symptoms. The doctor can then decide whether the patient needs a visit.

There are so many different viruses for influenza that it is difficult to be protected against it. There may be an **epidemic**, which is an outbreak of the same disease in towns, cities and regions. Sometimes there is a **pandemic**, in which whole continents are infected with a type of influenza. Once the 'flu virus settles in the respiratory passages, it multiplies by the millions! It spreads far more quickly than the common cold. The World Health Organization has an alert-system to track down new viruses which may cause epidemics or pandemics. There have been two pandemics of influenza this century, which have caused much suffering and death.

The air we breathe

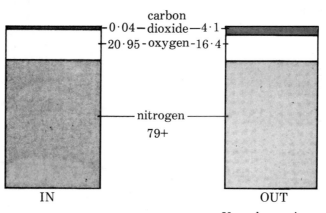

carbon

0·04 — dioxide — 4·1
20·95 - oxygen - 16·4

nitrogen
79+

IN

OUT

Usually quite dry air
Usually quite cool air
Usually quite dirty air

Very damp air
Very warm air, at body temperature
Cleaner air, unless we have an infection and breathe out germs.

Fig 5.12 The air we breathe

57

Nitrogen
Our lungs cannot use this gas. It is breathed in as far as the alveoli and then breathed out again.
Oxygen
This is the gas we must have to live. It is curious to see that our lungs can only use a part of the oxygen with each breath.
Carbon dioxide
There is very little carbon dioxide in the air we breathe in. But we breathe out quite a lot. We are like machines, able to make waste gas.

a

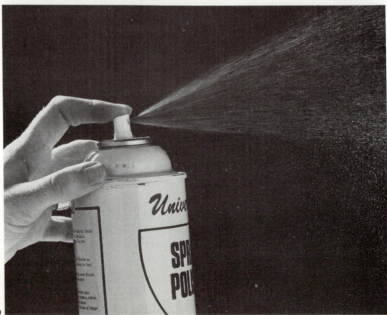

Fig 5.13 a Pollution from industry
b Pollution at home

b

Dry/wet air

The air we breathe in is usually dry, unless we live in a very humid climate. It is always soaked with moisture when we breathe out. (Why do we hold a mirror against the nose or mouth of a person to find out if he is still breathing?)

Cool/warm air

Air breathed in is usually cool. Air breathed out has been warmed by the blood capillaries in the respiratory passages. It is breathed out at body temperature.

Dirty/clean air

In the country, the hills and mountains, air is fairly clean. Air in the towns, cities and suburbs is polluted. As some of the things we breathe in are trapped by the mucus lining, the air we breathe out is cleaner than the air we breathe in. (Unless we have some respiratory infection.)

Pollution of the air

Whether a country is rich or poor, wherever there are a lot of people near one another making and using machinery, air pollution happens. Cities and large towns with factories suffer from heavy **air pollution**. In Britain, during the winter of 1952/3, the huge amounts of smoke from factories and home fires mixed with the fog. Thick, horrible air called **smog** hung over the towns, cutting out the light, the fresh air and the sunshine. In London alone during that winter, 4000 people died from the smog.

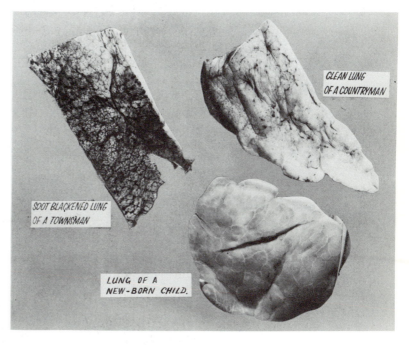

Fig 5.14

At last, the Government passed two Clean Air Acts, one in 1956 and the other in 1968. These acts made it an offence for factories or private homes to produce dark smoke from their chimneys. In the most polluted areas in Britain, smokeless fuels had to be used.

The cities and towns are much cleaner now. But with more industry, more factories, more cars, more aeroplanes, the pollution level will probably rise again. Large towns all over the world have this problem.

Air pollution is very dangerous. Our breathing system can only cope with small amounts of dust and dirt. Slowly, the soot and filth build up in the lungs and breathing tubes, making scratchy, gritty patches on the delicate mucus membrane. Germs can grow easily on these raw places. The lungs become scarred and diseased by bronchitis. When a person keeps getting attacks of the same disease, it is called a **chronic disease**. You may know someone who suffers from chronic bronchitis, as it has been called the British disease. Chronic bronchitis causes 25 000 to 30 000 deaths a year and costs the country £72 million in health and Social Security bills. No wonder we are all anxious to clean up the air we breathe.

Self pollution (smoking)

It wasn't until quite recently that smoking was found to be dangerous. Before this was known, people began to smoke when they were young, they liked it, they became **addicted** or 'hooked', and didn't bother too much when they found they couldn't give up the habit. When they had bad coughing fits, chronic bronchitis, serious shortness of breath, weak hearts, damaged arteries and lung cancer, they thought it was from other causes. Now it is known that smoking not only damages the lungs, but also damages the heart and arteries.

Until these facts were known, doctors were among the heaviest group of smokers. Doctors have been able to give up the smoking habit because they see their patients struggling to breathe as their hearts and lungs weaken. They have to watch them die. But it is not so easy for other people to stop smoking. The habit is so powerful, we become so addicted to cigarettes, that it is very difficult to break the habit and stop the craving. We do not have the help that doctors have. We do not *see* the slow but steady damage being done in our hearts and lungs. We see people who look quite fit and healthy smoking. We don't really believe that one day it will be our turn.

FACTS ABOUT SMOKING DAMAGE

Certain chemicals in the tobacco smoke cause cancer in the lung tissue. Lung cancer destroys the lungs and is a killer.

Carbon monoxide, a gas in tobacco smoke, mixes with the haemoglobin in the red blood cells. This makes the blood less efficient at carrying oxygen. The arteries round the heart are weakened and may cause a heart attack.

Babies born to mothers who smoke may be less healthy than babies born to non-smoking mothers. This is because the blood supply from the mother to her unborn baby is affected by the tobacco smoke.

Smoking damages the working of the cilia so that they cannot push the mucus which has trapped the dirt particles up and out of the lungs and breathing tubes. The phlegm gets heavier and stickier. It drops down into the breathing spaces and collects into horrid pools, causing infection and disease.

Smokers cough to try to bring up the heavy phlegm. Long fits of coughing damage the delicate linings of the breathing tubes and lungs. There is less space for oxygen to pass into the bloodstream so the smoker gets 'short of breath'. He has attacks of bronchitis, becomes a chronic bronchitic, and the strain on his heart and lungs becomes too great.

These are some of the more serious facts known about smoking and health. Lots of research is going on into the dangers of smoking and it is thought many more dangers will be discovered.

WHY DO PEOPLE SMOKE?

Because lots of other people do. And because we believe it helps us to be more grown-up. It's really as simple as that. Someone who has never smoked doesn't have a craving for cigarettes, so the reasons for starting to smoke are quite simple. Once you start smoking, you are 'hooked'.

Smoking is a drug which affects the whole body. The smoker begins to enjoy smoking; it calms the nerves, is something to do in awkward moments and satisfies the cravings for tobacco smoke. It is a **comfort habit** which is rather the same as the dummy, the doll, the teddy, or the soft cloth small children suck and hold and keep close to them when they are little and frightened. Smoking is a grown-up comfort habit but with all the extra problems of addiction and ill-health.

BEGINNING TO SMOKE

Smoking the first cigarette is not much different from trying on lipstick or tasting beer. Most children try their first cigarettes at any age from 3 to 10. Can you remember your first? The second cigarettes *do* matter. We are older and want to belong to the world of older people. So we smoke, which is one of the things they do. Studies have shown that a teenager who smokes two or three cigarettes has a 70 per cent chance of smoking for the rest of his or her life. It is hardly believable that we become hooked, dependent and addicted so quickly, but it is absolutely true. The smoking habit gets a powerful hold over us incredibly quickly. And the earlier we start smoking, the bigger the risk of early disease and death.

Fig 5.15

WHAT STARTS PEOPLE SMOKING WHEN THEY KNOW IT IS SO
DANGEROUS?

We don't know the answer to this one. There is no point
asking older smokers why they started because they did not
know it was bad for their health. But now we do know it is
very dangerous. And yet people still begin to smoke.

Smoking goes on anywhere the young smoker isn't likely
to be caught – in toilets, bedrooms, anywhere private. Do
you think smoking in secret makes it more exciting, more
daring? Can you think of any answers as to why people
start nowadays?

IF YOU ARE ALREADY A SMOKER

1. Don't offer cigarettes to your friends. 'Have a cancer
    tube' or 'Have a sick-stick' is not exactly friendly.
2. Don't smoke where other people are not smoking. They
    have to breathe in your cigarette smoke and it can
    affect their lungs.
3. Don't smoke if there is a baby or child in the room.
    Their lungs are easily polluted by your tobacco smoke.
4. Don't smoke in front of younger people who admire
    you. They may want to copy what you do and you could
    be responsible for damaging their health.
5. Go to a Smoking Clinic where you can get all the help
    you need to break yourself of the habit.
6. It is very important that a girl gives up smoking. Later
    on, if her baby is born less healthy, she will feel
    dreadfully guilty she didn't break the habit when she
    was younger.

The function of the breathing system

The function of the breathing system is to get air in and out of the lungs. On average, we breathe in and out about 15 times a minute. Children and women breathe slightly more quickly than men and we all breathe much more rapidly when we take exercise. We can stop ourselves breathing for a short time, but really all breathing goes on without our thinking about it. It is controlled by part of our brain.

HOW WE BREATHE

Our body is divided into two parts or **cavities**, the **thorax** or chest, and the **abdomen**, which contains the stomach and other organs. Breathing only happens in the thorax. The two main organs in the thorax are the lungs and the heart. They are protected by the pleurae and the rib-cage, which make up the chest wall. The **diaphragm** is a sheet of muscle between the two cavities. 'Dia' means across and 'phragm' means fence, so the diaphragm is like a fence across the body, separating the thorax from the abdomen.

Both the diaphragm and the rib-cage help us to breathe.

Breathing in

The **intercostal muscles** between the ribs pull the ribs upwards and outwards. This makes the lungs stretch and expand so air is sucked and dragged down the nose and trachea into the lungs.

At the same time the muscles of the diaphragm contract. pulling it down and flat. This makes the lungs stretch and expand so air is sucked or dragged down the nose and trachea into the lungs.

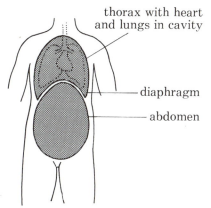

thorax with heart and lungs in cavity

diaphragm

abdomen

Fig 5.16 Body cavities

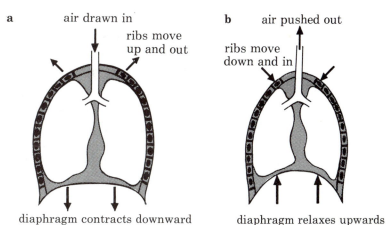

a   air drawn in   ribs move up and out   **b**   air pushed out   ribs move down and in

diaphragm contracts downward   diaphragm relaxes upwards

Fig 5.17 a Breathing in
       b Breathing out

Breathing out

The rib cage muscles relax so that the ribs move downwards and inwards. This squashes the lungs, forcing the air out of them.

At the same time the muscles of the diaphragm relax and the diaphragm curves upwards again. This squashes the lungs, forcing the air out of them.

The important things to remember are:

a.   when we breathe in we make room for air to be sucked in from the outside atmosphere;

b.   when we breathe out, we squash up the space in the lungs so the air has to be forced out.

If you can, work out how air is drawn into bellows, as it is rather the same in the lungs.

Air breathed in is called **inspired air**, and air breathed out is called **expired air.**

HOW MUCH AIR DO WE BREATHE?

You can find this out for yourself on page 66.

The amount of air we normally breathe is called **tidal air**. Like the tide, it flows in and out quite regularly. There is about 500 cubic centimetres of it. But if you take a breath, an ordinary tidal one, and then try breathing in more air, you will find you can do it quite easily. If you breathe out normally, and then try to breathe out more air, you can also do this. This extra air we can take in and out is called **supplementary air**. Supplementary air is about three times the amount of tidal air. And then there is always some air in the lungs which we cannot breathe out, no matter how hard we try. This is called **residual air** and is about twice the amount of tidal air. (When someone is hit very hard in the 'bread basket', just below the diaphragm, they feel winded. This is because some of the residual air may be knocked out of the lungs.) How much air we can get into and out of our lungs is called our **vital capacity**.

Breathing quietly, tidal breathing, and breathing deeply during exercise or from fright or strong emotions, is called **external respiration** or gaseous exchange. External respiration is the getting of air from the outside into our lungs and the returning the waste air from our lungs to the outside.

Internal respiration

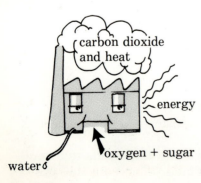

Fig 5.18

WHY IS OXYGEN SO IMPORTANT TO US?

The cells of our body have a lot of work to do. As well as doing their own special work, they have to keep healthy, make new cells for growing, repair old worn-out cells and many other things. To do work, they must have *energy*. How do they get the energy they need?

Look back to page 4 and study how things are passed in and out of the cells. The two things which pass into the cells to provide energy are oxygen and food. Oxygen is taken to the cell in the haemoglobin of the red blood cells. Food, broken down into simple sugars, is taken to the cells in the bloodstream.

When the oxygen and sugar are in the cell, the oxygen acts on the sugar as if it were burning it, setting free the energy stored in it. The energy is then used for the work of the cell. Think of it like this:

As the oxygen is acting on the sugar, heat, moisture and a waste gas, carbon dioxide, are made.

We can think of the work of the cell as rather like a tiny factory. Oxygen and sugar are delivered to the factory. Energy is the product which is set free from the work going on in the factory. So:

$$\text{SUGAR} + \text{OXYGEN} = \text{ENERGY} + \text{HEAT} + \text{MOISTURE} + \text{CARBON DIOXIDE}$$

The cell passes back the three products into the bloodstream. Heat is taken around the body to keep it warm, while the moisture and carbon dioxide are taken to the lungs and passed out when we breathe out.

All factories produce waste; the cell produces waste gas, carbon dioxide, which is taken back to the lungs in the plasma of the blood. It is strange to know we actually make a gas which is poisonous to us. If we did not breathe out, the carbon dioxide level in the blood would rise. Too much carbon dioxide in the blood causes urgent messages to be sent to the brain for us to breathe out. The need to get rid of carbon dioxide controls our breathing rate. If we hold our breath, we feel a need which we can't control to start breathing again. This is because of the build-up of carbon dioxide in the blood.

When we do a lot of exercise, both the lungs and the heart work far more quickly. Why do you think this is important?

CUTTING OFF THE AIR SUPPLY

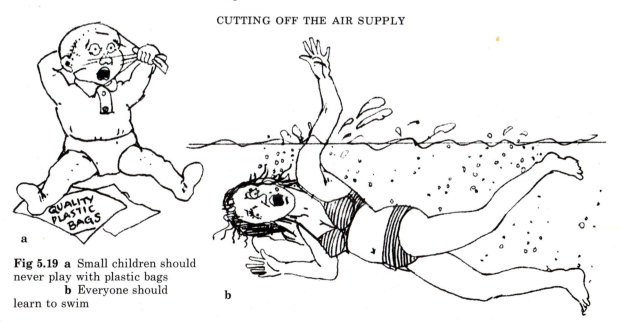

Fig 5.19 a Small children should never play with plastic bags
   b Everyone should learn to swim

Artificial respiration

We cannot live without oxygen for more than 5 minutes, (page 143), so speed is vital to save a person's life if they have stopped breathing. Artificial respiration must be started at once and it must not be stopped till a doctor arrives, or an hour has passed without the person beginning to breathe. There are many different methods of artificial respiration. They all help to force air in and out of the lungs. You must be properly trained before you practise. And you must not practise on your friends unless you are well qualified. Not knowing what you are doing can be very dangerous.

Fig 5.20 Artificial respiration

More facts about breathing

Breathing is stopped for the short time when we swallow.

To laugh, we breathe in deeply, tighten the vocal cords, then let our breath out in short gasps. Crying is almost the same.

Yawning and sighing are huge gulps of air taken in or let out slowly. Hiccoughs, or hiccups, are quick breaths ending with a click as the vocal cords suddenly close. They are caused by a nerve twitching in the diaphragm.

An unborn baby doesn't 'breathe' through its lungs. The oxygen in the mother's blood is taken to the baby's cells so that internal respiration can be carried out. A baby takes its first breath of *air* when it is born.

An unconscious person must not be given a drink. They are unable to swallow and the water will go straight down the trachea into the lungs.

The vital capacity of your lungs can be increased by practising deep breathing. This is very useful for swimmers.

Questions and things to do

Find your vital capacity (Figure 5.21)
Practise deep breathing a few times, then blow as hard as you can, and for as long as you can, down the short tube. Measure the amount of water you have blown out. Do this a few times. The largest amount of water you blow out is the one nearest to your vital capacity.
Find your tidal breathing (Figure 5.22)
Practise breathing in and out of your mouth, holding your nose. Then, as normally as possible, breathe in the air from the bottle through the tube. After breathing in at tidal rate, take out the tube and put a cork in the neck of the

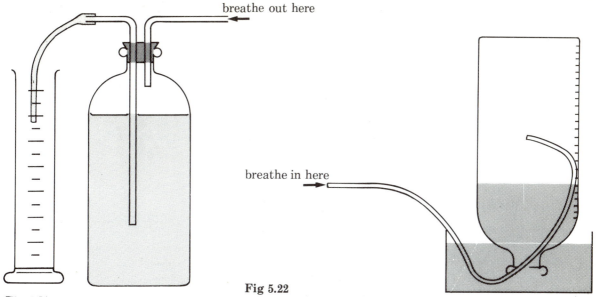

breathe out here

Fig 5.21

breathe in here

Fig 5.22

bottle. Measure the amount of water. This amount, or volume, is your tidal breath.

To show the air you breathe out has more carbon dioxide than the air you breathe in (Figure 5.23)

The liquid at the bottom of the flasks is **lime water**. Lime water turns cloudy when carbon dioxide passes through it. Breathe in and out through the Y-tube until you see a change in the lime water. Write down what you think has happened.

Diagrams to show diaphragm breathing (Figure 5.24)

Draw and explain in your own words what happens when you pull the string down and when you let it go.

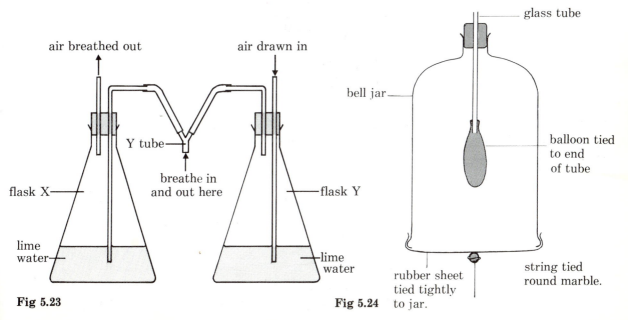

air breathed out

air drawn in

Y tube

breathe in and out here

flask X

flask Y

lime water

lime water

Fig 5.23

glass tube

bell jar

balloon tied to end of tube

rubber sheet tied tightly to jar.

string tied round marble.

Fig 5.24

1. Why is it healthier to breathe through your nose?
2. Copy the picture of how swallowing takes place (page 51). Explain what happens in your own words.
3. Explain carefully what is meant by 'gaseous exchange' or 'external respiration'.
4. Make a list to show the passage of air out of the lungs.
5. What are cilia and what is their function?
6. How should we get rid of phlegm?
7. What things would you do to try not to catch a cold?
8. Explain the difference between an epidemic and a pandemic.
9. Copy out the differences between inspired and expired air. Learn them.
10. Do a project on any one form of pollution.
11. Do a project on any one disease of the breathing system which interests you. (You can also choose from whooping cough, tuberculosis, asthma or diphtheria.)
12. What is meant by 'self pollution'? Do a project on smoking and health. Write to Action on Smoking and Health or the Health Education Council for information.
13. Have a class debate on smoking.
14. Write out and learn the facts about smoking damage.
15. Visit your nearest smoking clinic to see the work being done.
16. Design a really exciting advertisement to persuade people (*a*) not to start smoking or (*b*) to give up if they have already started.
17. Explain carefully, with diagrams, how air is forced in and out of the lungs.
18. Try to explain, in a simple way, how energy is set free in our cells.
19. Internal or tissue respiration is sometimes written like this:

$$C_6H_{12}O_6 + 6O_2 = 6CO_2 + 6H_2O + \text{ENERGY}$$

sugars   oxygen   carbon dioxide   water

Can you explain why?
20. Why do we pant when we do a lot of exercise?
21. With a stop watch, measure your breathing rate:

    a. when you are sitting quietly;
    b. after 2 minutes' exercise;
    c. after 4 minutes' exercise;
    d. 1 minute after you have stopped exercising;
    e. 2 minutes after you have stopped exercising.

Write down the results and explain why you go on panting after you have stopped exercising.
22. Central heating in homes, schools and work places can make the air we breathe too dry. This may affect the mucus lining of the respiratory system. Do a project on heating and ventilation and how they affect our health.

23. What is 'air conditioning'? Find out all you can about it and make a list of the countries and places where it is used.
24. Find out if your local Safety Officer can visit your school to give a demonstration of artificial respiration.
25. On the outline of a skater, draw the lungs in their correct position.

Chapter Six

Foods

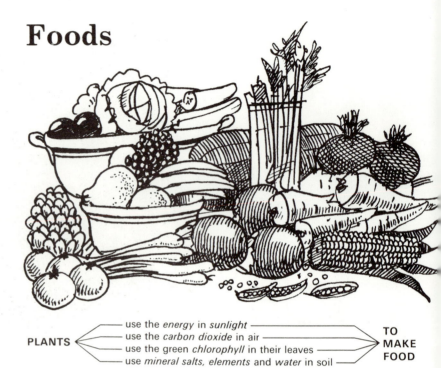

Fig 6.1

PLANTS
— use the *energy* in *sunlight* —
— use the *carbon dioxide* in air —
— use the green *chlorophyll* in their leaves —
— use *mineral salts, elements* and *water* in soil —
TO MAKE FOOD

Food is made for us by the plant world

Animals and humans cannot make their own food. Only plants can do this extremely important thing. They are able to make food in sunlight by a process called **photosynthesis**. Animals get their food by:

1. *Eating plants.* Cows, sheep, goats, elephants, worms, butterflies, kangaroos, and tiny fish are some animals which get their food from plants. Plant eaters are called **herbivores**.
2. *Eating animals.* Sharks, lions, vultures, penguins, frogs, snakes, spiders and fleas are some animals which get their food from other animals. Animal eaters are called **carnivores.**
3. *Eating both plants and animals.* Many animals like to eat both kinds of food. Write out a list of as many as you can think of. Animals which get their food from both plants *and* animals are called **omnivores.**

Humans are omnivores. We eat plants and we eat animals. We also eat things which some animals produce, such as milk and eggs. But the animals we eat have got their food from either (a) plants they have eaten, or (b) other animals, which in their turn, ate plants. So all our food comes from plants, even when we eat animals or animal produce. We can trace everything we eat back to plants.

$$GRASS \rightarrow COW \rightarrow (BUTTER) \rightarrow MAN$$

$$TINY\ PLANTS \rightarrow TINY\ FISH \rightarrow BIGGER\ FISH \rightarrow MAN$$

These are simple food chains.

Fig 6.2 A Japanese family eating raw fish

Why we need food

1. We need food for energy. We have already studied how simple sugars and oxygen are taken to the cells and how the energy in the sugar is set free by the oxygen. This energy is then used by the body for all the different types of work it has to do.
2. We also need food for growth, for body-building, for replacing old cells, for mending tissue, for warmth, for keeping healthy, for storage and for many other functions.

Food groups

Our bodies need different foods for different purposes. The different food groups are: proteins, carbohydrates, fats, water, vitamins, and mineral salts and elements.

Why do you think plant foods are so much cheaper than animal foods?

How much do we need each day:
Carbohydrates and fats are energy foods. The amount of

energy foods we need daily depends on many different things. Our age, sex, the work we do, how active we are, our size and the climate we live in are some of the different things. When you study food charts, remember they are worked out on *averages*. We may need a little more or a little less than is written on the charts. The energy we get from food is measured in **kilojoules**. It used to be measured in **Calories**. Both measurements will be used in these food charts.

Study the charts very carefully and make notes on all the interesting differences.

Proteins are building foods. The amount of protein foods we need daily is about 1 gram of protein for every 1 kilogram of body weight.

Proteins	Carbohydrates	Fats
Animal proteins Plant proteins	Starches Sugars	Animal fats Plant fats or oils.
Needed in diet for:		
Growth, body-building, replacing old cells. Can be used for energy when needed.	Energy. Food storage Heat energy.	Energy, Heat energy and insulation. Food storage.
Special needs:		
Women who are pregnant or breast-feeding. Babies, children and adolescents.	People doing heavy physical work or a lot of extra sport.	People living in very cold countries eat a lot of fats.
Deficiency (not enough of:)		
Very serious in babies and growing children. The disease is called Kwashiorkor.	Serious loss of weight. Not enough energy to lead a healthy life.	Poor growth and skin. Loss of weight. Not enough energy or heat energy.
Cost:		
Animal protein is very expensive but it is the best protein. Plant protein is much cheaper but may lack some important things.	Cheapest foods because they come directly from plants.	Animal fats are usually far more expensive than plant oils. Fats have more than twice the energy value of carbohydrates.

Protein, Fat, Carbohydrate in g. per 100g. Tr = Trace

Food	Protein	Fat	Carbohydrate	Kilocalories per 100g	Kilojoules per 100g
Apple	Tr	Tr	10	36	151
Orange	1	Tr	6	27	113
Beans, Runner	1	Tr	1	7	29
Carrots, boiled	1	Tr	4	19	80
Peanuts, Kernels	28	49	9	600	2520
Potatoes, boiled	1	Tr	20	80	336
Bread, white	8	1	53	240	1008
Butter	Tr	85	Tr	790	3318
Sugar, white	Tr	Tr	105	390	1638
Milk, fresh whole	3	4	5	66	277
Cheese, Cheddar	25	35	Tr	430	1806
Eggs, fresh whole	12	12	Tr	160	672
Sardines, canned	20	23	Tr	290	1218
Cod, fried	21	5	3	140	588
Pork, roast	25	23	Tr	320	1344
Chicken, roast	30	7	Tr	190	798

Table adapted from Medical Research Council special report 297, H.M.S.O. publications.

WATER

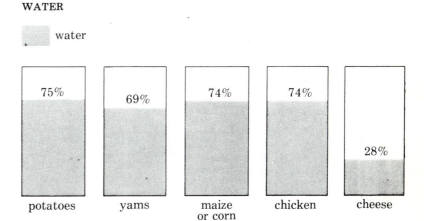

Fig 6.3 The water you eat

We get about *half* our water from the food we eat. A potato has 75 per cent water, a yam 69 per cent, corn 79 per cent, chicken 74 per cent, cheese 28 per cent.

Needed in your diet for:
Without any water at all, we would die in a few days. All body cells, tissues and fluids are made up of lots of water. Cell protoplasm is about 75 per cent water. Nearly three-quarters of our body weight is water. Water is essential for transport, taking things all around the body. Water helps in chemical changes which happen in our body. It helps to remove waste products. It helps to keep our body temperature even.

Special needs:
How much water we need is carefully controlled, so that the amount of water in our body is kept fairly even. Each day we lose between 2 and 3 litres of water, from waste urine and food, from the skin in sweat and from the lungs in breathing out. Each day we *must* put back the 2 to 3 litres we have lost. Some people lose more water than others. An Indian working in the sun may lose up to 11 litres a day, and a miner can lose 7 litres during one shift underground. A person who has a fever needs lots of extra fluid because he sweats so much. We sweat about 30 cubic centimetres every hour. When we exercise, we sweat a great deal more. People doing heavy physical work or a lot of extra sport need to take in more water. People living in hot, tropical climates sweat very heavily to keep cool, and must be sure to replace the water they lose.

Deficiency:
Any lack of water upsets the balance between how much we lose and how much we take in. A serious lack of water leads to **dehydration**; the body is dried out and cannot work properly. People with **diarrhoea** ('runs'), especially babies, often suffer from dehydration. People with fevers or in tropical climates may also become dehydrated. Dehydration is extremely dangerous. Water must be given at once.

Cost:
People in towns must pay water rates. In dry countries, water is scarce and is much more expensive. Water, like air, can be polluted. Water supply, and how it is cleaned, is studied in Chapter Nineteen.

How much do we need each day?:
We need between 2 and 3 litres, though this amount will depend on many different things. Remember, we get about half our water from food.

VITAMINS
Vitamins are complicated **chemicals**. They do not have any energy value. They are essential for the chemical changes which go on in our body. We only need very small amounts of vitamins in our diet, but without these small amounts the body cannot work properly and we may get serious diseases. A disease caused by a lack of a vitamin is called a vitamin **deficiency disease**. Most deficiency diseases can be cured by putting the vitamin back into the diet. Vitamins are found in all the food we eat, except sugar which has no vitamins. A *mixed diet*, eating plenty of different foods, will usually give us all the vitamins we need.

MINERAL SALTS

Mineral salts are chemicals containing elements needed in our diet. They are not used for setting free energy. They are essential: for making bones and teeth, for the growth of body cells, especially blood cells, and as ingredients of different body fluids. We need even smaller amounts of them than we do of vitamins. Babies and small children must have them for growth. Adults need them to replace those lost in waste urine and food, and in sweat. A lack of any of these elements is also called a deficiency disease.

Iron is needed for haemoglobin in the red blood cells.

Calcium, magnesium and *phosphorus* are needed for bones and teeth.

Calcium is also needed for blood clotting and the proper working of nerves.

Iodine is needed for the **thyroid glands** (page 161), which control growth.

VITAMINS IN THE BODY			
Vitamins	Best foods	Essential for	Deficiency disease
A (Can be stored in the liver)	Fish-liver oils Sardines. Liver, Milk and butter Fresh green veg.	Growth. Health of eyes. Protection from infections	Night blindness. Likely to catch more diseases. *Xerophthalmia,* an eye infection.
B₁ (Cannot be stored)	Yeast, wheat-germ, egg-yolk, liver, soya beans	Growth. Proper working of heart. nerves, muscles and digestive system.	*Beri-beri* (meaning weakness) Tiredness, loss of appetite. Weakness.
B₂ and other B vitamins	Yeast, milk, cheese, kidneys, peanuts, meat, fowl.	Growth, health of skin. Proper working of body	*Pellagra,* upset working of the mind. Skin infections. Weakness.
C (lost by over-cooking or storing)	Citrus fruits, nuts, tomatoes, onions, pineapples, berries, green vegetables.	Healthy teeth, gums and mouth. Healing of wounds Healthy bones and blood vessels.	*Scurvy.* Loose teeth, bleeding gums, bruise easily, aching joints.
D (Can be made in skin in sunlight) and stored.	Fish-liver oils Eggs, liver, milk.	Vital for body to use calcium and phosphate in making healthy bones and teeth.	*Rickets,* children's bones are too soft, legs and body deformed. Bone softening in pregnancy and old age. Bad teeth.
E	Green vegetables Wheat germ, milk.	Thought to be helpful for sex organs and having a baby.	This is not yet proved.
K. (stored in liver)	Dark-green leafy veg. (found in most food.)	Helps in the quick clotting of blood. Helps in liver function.	Not usually absent from diets.

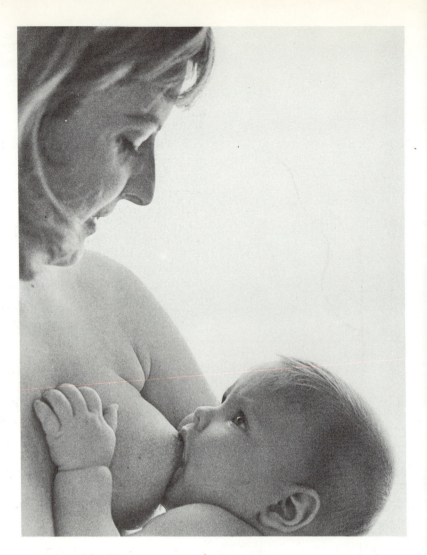

Fig 6.4 Breast feeding

Fluorine as fluoride, helps to prevent teeth decaying. *Sodium* and *potassium* are needed in nearly all the body cells, blood fluid and in nerve tissue.

These are a few of the mineral salts and elements needed by the body. They are in most mixed diets, though there may be a shortage of calcium (found in cheese and milk), iron (found in liver and red meat) and iodine (found in fish and added to salt). A shortage of calcium may lead to rickets, a shortage of iron may lead to anaemia, and a shortage of iodine may lead to growth being too fast or too slow.

Milk

All babies, black, white, brown, yellow or red, develop best on breast feeding. Human milk is an ideal food as it has proteins, carbohydrates, fats, vitamins, mineral salts and water. The baby is born with enough stored iron to keep him healthy till he is ready for solid foods. Babies may be given vitamin C in a fruit drink. Vitamins A and D may also be given to babies over 1-month-old by giving them cod-liver oil. Breast milk is better than cow's milk because it exactly matches human needs and is easier to digest. So breast feeding is better than bottle feeding. It is now thought fashionable to bottle feed, which is a pity for the baby (see Chapter Fourteen).

Roughage

This word is the name given to the tough part of food we cannot digest. It is the plant fibre in such things as brown bread, tomato skins, celery, string beans and other fruit and vegetables. These fibres in our diet help the food to be moved down the waste tube. The waste food, **faeces,** becomes solid and is easily got rid of when we go to the lavatory. **Constipation** may be caused by lack of roughage in the diet.

A balanced diet

A balanced diet means we must eat enough to get:
> *kilojoules* or Calories of energy;
> *proteins* for body-building and replacement of cells;
> *carbohydrates* for energy;
> *fats* for energy and heat;
> *vitamins* for healthy working of the body;
> *mineral salts* for healthy working of the body;
> *water* for transport and all important body functions;
> *roughage* for helping to get rid of waste food.

Some things we eat are rich in certain foods but may be poor in or have none of the other types of food.
Study the food charts carefully. Work out which foods are best for (*a*) a rice planter working in the hot sun, (*b*) a pregnant woman, (*c*) a growing child, and (*d*) a person who is underweight.

Starvation

Starvation is not getting enough kilojoules or Calories of energy to keep you alive. A person who is starving loses his body weight, the skin looks old, the heart beat slows down, the muscles wither, and breathing gets difficult. Fluid collects in the legs and abdomen. This is why the pictures we see of starving children show them with swollen bellies.

About half the people in the world, mainly in parts of Asia, South America and Indonesia, do not get enough kilojoules to keep them healthy and active.

Malnutrition

'Mal' means bad or wrong. People suffering from **malnutrition** eat too much of the wrong food, or not enough of the right foods, or not enough of any foods.

KWASHIORKOR
This is a deficiency disease caused by not enough protein in the diet. Animal proteins cost more than other foods, so children from very poor families in the West may suffer

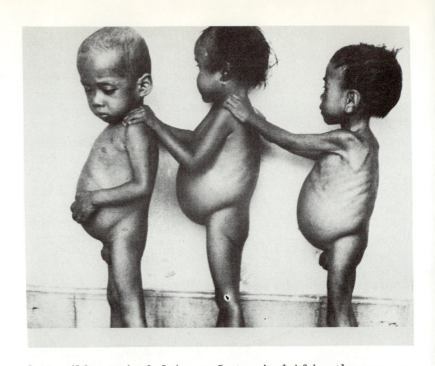

Fig 6.5

from mild protein deficiency. In tropical Africa there seems to be no shortage of food. But the word 'Kwashiorkor' comes from a language spoken in Ghana and it means 'the sickness that comes when a new baby is born'. Babies are breast-fed until the mother has her next child. Then the first baby is fed on fruits and vegetables, maize, yams and bananas. There is not enough protein in these foods for growth and body-building. The child becomes seriously ill and may die before he or she is 5 years old. Lack of protein and lack of carbohydrate often go together. Many, many babies and children suffer from protein-calorie malnutrition, P.C.M.

Malnutrition of any kind is a serious problem. Not only do we suffer from the deficiency disease itself, we also become generally unfit. We are then not able to fight off germs, so we are far more likely to get all sorts of other diseases. Malnutrition makes us less *resistant* to disease.

Another very different kind of malnutrition is obesity.

OBESITY

This is being too fat, being overweight. We become obese when we eat more food than we use up in energy. The extra food is changed into fat and stored in **adipose tissue** which is under the skin and around most of the organs. The sugar in sweetened food is probably the biggest cause of obesity in the Western world. Over-feeding a baby or child is 'a cruel kindness' because a fat baby is likely to grow into a fat adult. Fat people are not usually as healthy as slim people. There is a link between fatness and a shorter life, diseases of the arteries, liver and kidneys, high blood

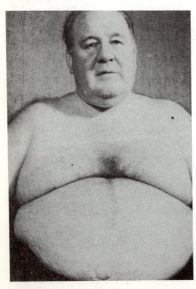

Fig 6.6

pressure, diabetes and many other illnesses. It is shocking to realize that people in some parts of the world are actually starving to death while people in other parts are getting ill through eating too much.

World food problems

SOME CAUSES

Natural problems. Drought, floods, high winds, pests, etc., ruin the crops.

Soil. This is often poor, or over-worked, lacking in mineral elements and salts.

Sickness. People have not enough strength to produce their own food.

Farming methods. Old farming methods don't produce enough food.

Population. A huge increase in the number of people to be fed.

Information. Not knowing enough about the right foods needed.

a

b

Fig 6.7 a Drought
b Old farming methods

THINGS BEING DONE TO HELP

Natural problems. Preventing soil erosion, irrigation, dams, controlling pests, etc.

Soil. Using fertilizers to enrich the soil.

Sickness. Prevention by hygiene of food and water, and destruction of disease-carrying insects.

Farming methods. Modern farming methods to produce more food, i.e. **intensive farming.**

Population. Free birth control to parents not wanting large families.

Information. Teaching the importance of a balanced diet. Scientists and nutritionists are working on finding new types of food which are cheap to produce and have all the necessary food groups for balanced diet.

Overseas aid. You can read more about overseas aid in Chapter Twenty-one.

How can we cure over-eating?

The answer to this problem is quite simple. Don't eat so much. Cut down on all foods with sugar in them. Puddings, cakes, biscuits, sweets, chocolate, and soft drinks are full of sugar. Sugar is only useful in our diet as an energy food. It has no other food value at all. It is also very bad for our teeth. If you are overweight, think of sugar as your greatest enemy. You may have a friend who eats twice as much as you do and never puts on any weight. This is because different people use up their food at different speeds. If you are fat, it is likely you are one of the people who use up food very slowly. You may think it unfair that you have a slower rate of **metabolism**, a slower rate of using food. But then, people are different and you may have many other things to be proud of.

Like many health problems, over-eating can start in early childhood. Children are given sweets and sweet foods to comfort them or as bribes, or rewards, or treats. Sweet foods are even used as threats to make a child behave. And so, as we grow up, sweet foods are thought of as comfort. When we are unhappy or anxious or nervous, we eat to comfort ourselves. This makes us get fatter. We then have an extra reason to be unhappy. We eat more to cope with the extra unhappy feelings. And so it goes on. A bad health habit like this, which goes round and round, is called a 'vicious circle'. It is very difficult to break. One very fat woman even had her jaws wired together to stop her eating! She could only sip nutrients through a small hole left for a straw.

But habits can be broken, though you have to be tough with yourself.

Remember:

a.   Sugar and sweet foods are very fattening.

b.   Starches are fattening.

Fig 6.8

c. Fried foods are fattening; boil or poach eggs, don't fry them.
d. Don't cut out fats completely, because they are essential. But eat less of them.
e. Don't cut down on proteins, vitamins, minerals or water. They are essential.
f. Never go on a 'starvation' diet. Your skin will look awful and you will be ill.
g. Alcoholic drinks and the 'mixes' to go with them are all fattening.
h. Take plenty of exercise for muscle tone and to prevent flabbiness.
i. Exercise alone will not slim you. You need to walk 60 kilometres to lose 400 grams of fat.
j. A slim baby is healthier than a fat, chubby baby.
k. Unhealthy eating habits often begin in early childhood.
l. Give a child love, not sweets.
m. Start the day with a nourishing breakfast.
n. Don't eat between meals.
o. It takes a lot of effort to lose weight. Don't despair. Keep on trying.

Vegetarians

A **vegetarian** is a person who will only eat plant foods. This may be because of his religious or his cultural beliefs. Or it may be that he dislikes the idea of eating 'flesh'. To keep healthy, a vegetarian must eat a wide variety of fruit and vegetables. There is a serious risk he will suffer from protein deficiency as the protein in plants does not contain all the essential nutrients that animal protein has. Nearly all plant protein is second class protein. If a vegetarian eats animal produce such as milk, butter, cheese and eggs he will get all the first class protein he needs.

Elderly people

Elderly people quite often suffer from deficiency diseases. This is because they may not feel strong enough to buy and cook proper meals or because they cannot chew very well with their false teeth. Without a balanced diet, their body temperature drops and they become ill very quickly. Elderly people do not need a lot of energy foods, but they must have fats for body warmth and they must have all the other nutrients.

Cooking

The advantages of cooking food before we eat it are:
a. The taste of the food may be improved.
b. The tough fibres we need for roughage are softened.
c. The food is easier to chew and swallow.
d. Many bacteria, viruses and worms in food will be killed.
e. The delicious smell of cooking food makes our mouth water and stirs up our appetite.

The disadvantages of cooking food are:

a. Meat which is badly cooked can be very tough to chew.
b. Poor cooks waste food by burning it or letting it boil away.
c. Vitamins B and C can be lost during cooking. These vitamins are water-soluble, unlike the others which are fat-soluble. Water-soluble vitamins can be destroyed by heat or lost in the cooking water. Fruit and vegetables should be plunged into the smallest amount of boiling water possible and cooked for the shortest possible time.

Some simple food tests

Collect very small portions of food such as, flour, egg-white, cheese, sugar, onion, potato, carrot and milk. Chop up and crush the solid food, put in separate test-tubes with a little water and shake well. The milk need not have water added.

TO TEST IF FOOD CONTAINS STARCH
Boil the solution and let it cool. When it is cold add a few drops, three or four, of **iodine solution**. Starch reacts with iodine and turns the solution dark blue.

TO TEST IF FOOD CONTAINS GLUCOSE (SIMPLE SUGAR)
Add **Benedict's solution** to the test-tube first, then boil it gently and very carefully. If the solution finally changes to a brick-red colour, then glucose is present.

TO TEST IF FOOD CONTAINS PROTEIN
Add **Millon's reagent** to the test-tube, then boil the solution. If it becomes a pinkish mass, then protein is present.

TO TEST IF FOOD CONTAINS FATS
Simply crush the chopped food between two pieces of clean paper. Grease will appear if fats are present. You may also add four drops of **Sudan Black** to the test-tube, shake it and then let it stand. Fat droplets will rise to the surface carrying the dye with them.

Questions and things to do

1. Look up and read the story of the discovery of vitamins.
2. Do a project on world food problems.
3. Find out all you can about intensive farming.
4. Draw a map of the world and mark in the areas which have (a) too much water, (b) not enough water. What is meant by **irrigation?**
5. The sea is a rich source of food. Find out all you can about different methods of fish farming.
6. Clubs such as 'Weight Watchers' are quite successful at helping people to slim. Visit a slimming club and write down (a) in what ways they are helpful and (b) in what ways they could help people more.

7. Make sure you know your food tests and can do them quite quickly.
8. Most schools are multi-racial, so you can find out about different foods eaten by different races. Do a project on the diet of people from one other country.
9. Write out your own favourite meal. From which food group does each part come? Is it a meal that will keep you fit and healthy or does it lack certain nutrients?
10. 'Our energy comes from the sun'. Is this true? Give reasons for your answer.
11. Learn the six food groups, why they are important for the body and which foods are rich in each group.
12. What is meant by 'a balanced diet'? Write out a day's menu for (*a*) a 6 year old, (*b*) a mother who is breast feeding, (*c*) an overweight 10 year old, and (*d*) an elderly person in a very cold country.
13. What is a deficiency disease? Do a project on one of the deficiency diseases mentioned.
14. Why is sugar sometimes called as 'dreadful a curse as tobacco'?
15. Why are animal foods so much more expensive than plant foods?
16. Some slimming diets are dangerous because they do not supply enough of the essential foods. Work out a slimming diet which is safe.

Chapter Seven

Food in the body

Fig 7.1

She is only partly right. We *are* made up of the foods we eat, but food doesn't travel inside us in little lumps. A lot of things have to happen to it before we can use it properly.

Imagine you have just swallowed the last mouthful of your favourite meal. It has disappeared inside you and that seems to be the end of it. But, in fact, that was just the beginning of what happens to food in our body. That delicious meal has to be broken down and changed in many ways before your body can use it. How is it that chips, sausages, beans, ice cream, whatever you eat, can be used to make blood, bones, muscle, energy, brain cells, hair? It is all quite amazing! So you won't be too surprised to see that the system which does most of this work, the **digestive system**, looks pretty amazing too.

The food canal

Study Figure 7.2 carefully. Notice that the food canal is one long tube. See that it is wide in some places, narrow in other places and that it bulges right out to form the stomach. Find the diaphragm. You will see that the

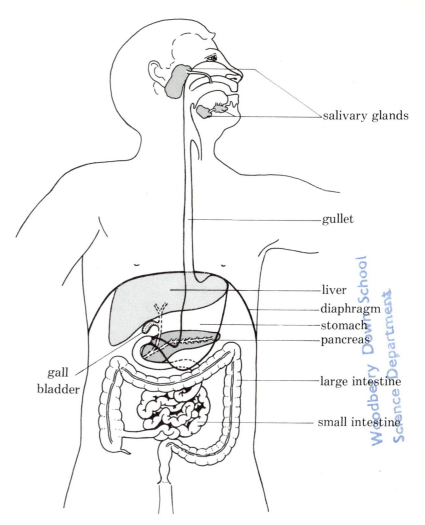

salivary glands

gullet

liver
diaphragm
stomach
pancreas

large intestine

small intestine

gall bladder

Fig 7.2 The food canal

stomach and most of the food canal are below it. This means that most of the digestive system is in the abdomen. The food canal is long, closely packed and coiled in the abdomen.

THE GLANDS OF THE CANAL
Study Figure 7.2 again, this time looking at the parts which are shaded. These are the glands which help in digestion. They make **enzymes** which help to break down food. Find the salivary glands, the liver and the pancreas. Notice the gall bladder. Each gland pours its enzymes into the food canal down small tubes called **ducts**. Find the tubes leading from each of the glands.

ENZYMES
Enzymes are special chemicals made in the body which help to bring about the chemical changes in the body. Digestive enzymes break down foods. They work best at body temperature and are destroyed by heat. Each enzyme can only do one sort of work on one type of food.

BREAKING DOWN FOOD

Food is made up of many **molecules**, which are tiny particles joined together. The first thing which has to be done is to break the molecules apart and change them into simpler things. They also have to be made **soluble** (page 4), to be dissolved in liquid. Breaking down foods and making them soluble is called **digestion**. When the food is digested, it has to leave the food canal and pass into the bloodstream. (Remember that one of the functions of blood is to carry digested food to all the cells and tissues of the body.) Once the food gets to the cells and tissues, it is used to produce energy (page 64) and it is used for growth, body building, repairing cells and so on. A part of the food we eat cannot be broken down or used by the body. Waste and undigested food are passed out of the body when we go to the lavatory.

The mouth

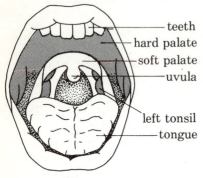

teeth
hard palate
soft palate
uvula

left tonsil
tongue

Fig 7.3 The mouth

We are going to begin our study of the digestive system at the beginning, at the mouth. Have a mirror beside you so you can check what you are reading about.

STRUCTURE

The **tongue** is made of voluntary muscle and lots of nerve endings. Have you ever scalded it or bitten it by accident? There are taste buds on the upper part (page 133). We use our tongue to talk, to taste our food, to keep food moving while we are chewing and to help us swallow.

The **hard palate** is the bony part of the roof of your mouth. We use it when we talk – try it – and to press food against while we are chewing.

The **soft palate** is the roof of the mouth near your throat. We use it to help us swallow.

The **tonsils** are on each side of the entrance to the throat.

The **uvula** is that curious bit of soft tissue which hangs down from the soft palate. We don't know what its function is.

The **teeth** were studied on pages 16–21. Make sure you know their structure, function and hygiene. Go back and re-learn them if necessary.

The **salivary glands** are in pairs on each side of your mouth. There are three pairs with ducts opening into your mouth. The top pair are called the **parotids** because this word means 'beside the ear'. When we have mumps, the parotid glands get infected and so our face swells out and we look 'mumpy'. If you lift up your tongue, you can see the other two pairs as slight bumps on your lower jaw and under the tongue.

FUNCTIONS OF THE MOUTH

We use our mouth for chewing and for the very first part of digestion.

Chewing

Our teeth tear off the food, then chop, grind and mince it into small pieces. The cheek muscles and tongue move the food around while our lower jaw moves up and down on its hinge joints. It used to be thought we should chew our food 32 times, once for each tooth. Most people chew far less than this. Which takes the most chewing: a piece of meat, a bite of bread or a bite of apple? Chewing gives our teeth the exercise they need. It also minces the food so the digestive juices in the mouth can work on it.

Digestion in the mouth

While we are chewing, the salivary glands pour **saliva**, a digestive juice, down the ducts and onto the food. Saliva is mostly water, with mucus, some mineral salts and an enzyme, **ptyalin**, dissolved in it. Saliva has lots of functions.

It softens and moistens the food we are chewing.

It dissolves the taste chemicals that are in food so we can taste them.

It keeps the mouth damp so we can talk and swallow comfortably.

The enzyme ptyalin begins to break down carbohydrates.

It helps control how much water we have in our bodies. We feel thirsty when we haven't enough water in our body because our saliva stops flowing. Our mouth feels uncomfortably dry, so we take a drink of water. We **secrete** about 1.5 litres of saliva during 24 hours. (A **secretion** is any fluid *made in* the body and *used in* the body.) Of course, we secrete far more saliva when we eat than at other times. If you think about sucking a sharp lemon, you can feel saliva pouring into your mouth. We also secrete more saliva when our feelings are upset. After a nasty shock our mouth may feel 'dry with fright' (page 144).

Fig 7.4 A person in shock must not be given anything to drink

Swallowing

When the food has been chewed and moistened, the tongue and cheek muscles roll it into a small wad called a **bolus**. The bolus is pushed to the back of the throat and swallowed. Can you write out exactly what happens when we swallow, and draw the diagram, from memory? If you can't, turn back to page 51 and re-learn it.

HYGIENE OF THE MOUTH

Regular cleaning of your teeth, a balanced diet with plenty of chewy foods, and rinsing your mouth with water will keep your mouth fresh and healthy. Bad breath, **halitosis**, is often caused by rotting teeth. Don't use a mouth spray to sweeten your breath. It only hides the bad smell while the decay gets worse. Go to your dentist. Other reasons for bad breath may be smoking, eating spicy food, drinking alcohol or having a cold or 'flu. **Mouth ulcers** are sores on the lining of the cheeks or gums. Check your diet and rinse your mouth out whenever you can. If they don't heal up after 2 weeks, go to your doctor for treatment.

The gullet (oesophagus)

The gullet is the first part of the food canal leading from your throat to your stomach. It is quite narrow and about 25 centimetres long. The gullet is lined with mucus membrane, just like the mouth and the rest of the food canal. When there is no food in the gullet, it is a flattened tube. So when a bolus of food is swallowed it doesn't just drop down into the stomach, it has to be pushed along the tube. All food throughout the food canal has to be pushed along by special muscles.

PERISTALSIS

Peristalsis is the word used to describe how food is pushed along the food canal. The whole of the food canal is made up of special muscles, some going *down* the canal and some going *around* the canal. They relax and contract in turn, making space for the bolus to move into and then squeezing it onwards from behind. The muscles work in wave-like movements called **peristaltic waves** right through the food canal. We have no control over peristalsis. The muscles which cause it are involuntary muscles.

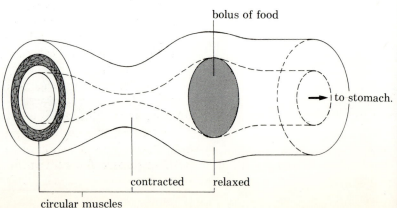

Fig 7.5 How peristaltic waves move food along the food canal

The stomach

STRUCTURE

The stomach is the widest part of the canal. It is like a large bag of muscle, just below the diaphragm. At each end is a special ring of muscle. (A ring of muscle in the body is called a **sphincter**.) The stomach can stretch to take in large amounts of food. When it is full, the sphincters close. When it is empty it is much smaller. Inside the stomach is a delicate lining which produces mucus, digestive juices and very small amounts of hydrochloric acid.

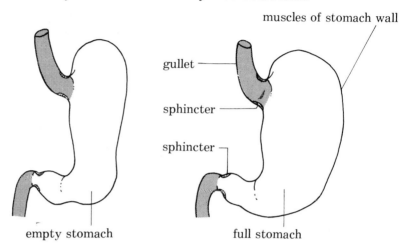

Fig 7.6 The stomach

FUNCTIONS

1 *It acts as a store for food.* A meal will stay in the stomach for 3 to 5 hours after you have eaten it. Without a stomach we would have to eat a little food every half an hour. Starchy foods leave the stomach first, meat and other protein foods leave fairly slowly, and fatty foods take longest of all.

2 *It churns the food till it is liquid.* While the food is being churned there is plenty of time for the digestive juices to do their work.

3 *It begins the breakdown of proteins.* The enzyme **pepsin** is one of the digestive juices made by the stomach wall. Pepsin begins the breakdown of proteins. The very small amounts of hydrochloric acid help pepsin to work. **Rennin** is the other digestive juice made in the stomach. Rennin curdles milk so that being half-solid it will stay longer in the stomach and there is more time for the protein in the milk to be broken down.

4 *It destroys many bacteria in food.* No matter how careful we are in preparing food, some bacteria may remain after cleaning and cooking. It is thought that the hydrochloric acid kills these bacteria.

Water, which doesn't have to be broken down, can pass through the stomach wall and into the bloodstream. So can alcohol, some mineral salts and simple sugars. Some drugs can also pass through. Aspirin, for example, will begin to work on your headache quite soon after you have taken it.

When the food has been in the stomach long enough, the inner sphincter opens and liquid food is squirted into the small intestine, a little at a time.

VOMITING

Being sick doesn't usually harm us though it feels horrid. Just before we vomit, our brain sends urgent messages to the diaphragm and the muscles of the abdomen. They contract strongly, forcing the partly digested food the wrong way out of the stomach and up the gullet. There are many different reasons for being sick. The most usual ones are nasty smells, over-excitement, disgust or too much of anything such as green apples, alcohol, rich food or tobacco.

Bumpy movements from being at sea or in a car may cause travel sickness. Some women in the first months of pregnancy get slightly sick. More serious causes are food poisoning, stomach disease and drug-taking. Babies and small children have very delicate stomachs and vomit quite easily. But if a baby is sick after each meal then there is likely to be something seriously wrong, and he or she must be taken to a doctor for treatment.

BELCHING

Sometimes we swallow small amounts of air when we eat. This makes us belch, especially after a big meal. Babies are helped to bring up 'wind' by being patted gently on the back. The air may travel through the food canal and come out at the other end, the anus. It smells unpleasant because of the gases made by bacteria in the last part of the canal.

Fig 7.7 Even too much of foods like these will make you sick

The small intestine (guts)

It is only called small because it is narrow. In fact, there is nothing else small about it as it is over 6 metres long, from the end of the stomach to the beginning of the large intestine. It needs to be this long because all the rest of digestion happens here and because the digested food is passed into the bloodstream here. The first part of the small intestine is called the **duodenum**.

The **pancreas** is just below the stomach. It is an important digestive gland as it produces three different enzymes to break down the three main food groups of proteins, carbohydrates and fats. It passes the enzymes down the pancreatic duct into the small intestine.

The **liver** is the largest gland in the body. It is also just below the diaphragm and almost surrounds the stomach. It produces a green, watery fluid called **bile**. Bile **emulsifies** fats. This means that bile splits the fat into tiny soapy droplets so the enzymes can work on it.

The **gall bladder** is below the liver and looks like a

small purse. It stores the bile and makes it stronger by removing some of the water. When food which is fatty enters the small intestine, the gall bladder contracts and squeezes the bile down the bile duct and onto the food.

The walls of the small intestine are lined with digestive glands which produce more enzymes to help in the work of breaking down the three types of food. It takes a long time to break down the foods completely. At last:

a.   Proteins are broken down into **amino acids**.
b.   Carbohydrates are broken down into simple **sugars**.
c.   Fats are broken down into **fatty acids** and **glycerol**.

Digestion is finished and the nutrients are ready to be passed into the bloodstream.

ABSORPTION IN THE SMALL INTESTINE

When little Johnny swallows a toy soldier, his mother knows it will reappear in his potty within the next few days. The enzymes from the digestive glands cannot break down the molecules of the toy. The toy cannot be absorbed, that is, it cannot be taken from the food canal into the bloodstream. Until something is actually passed into the bloodstream, it is not a part of our body. The toy soldier and waste food, though they have travelled right through us, cannot be thought of as once being a part of us.

HOW FOOD IS PASSED INTO THE BODY

Most digestion happens in the first half of the small intestine. Absorption, food being taken into the bloodstream, happens in the second part. Tiny, finger-like **villi**, millions of them, stick out into the intestine so the digested food has to flow over them. Each **villus** is made of a very fine membrane on the outside and tiny blood capillaries and a lymph capillary called a **lacteal** inside. Find them in Fig. 7.8.

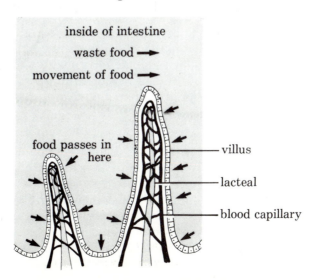

Fig 7.8 How food is passed into the blood stream

Can you imagine what it must look like? The walls of the small intestine move in peristaltic waves. The liquid digested food ripples forward over the villi. The millions of villi sway with the waves, the blood pulsing inside them as they absorb the food through their fine membranes.

Amino acids and simple sugars pass into the blood capillaries and are taken to join up with the **hepatic portal vein** (page 43). This large vein takes the food to the liver and then to all the cells and tissues of the body. Some fatty acids and glycerol pass into the blood capillaries of the villi but most of them pass into the lacteals. The lacteals join up with other lymph capillaries and the digested fats are taken by the lymphatic system to a large vein near the heart where they enter the bloodstream (page 45).

HOW FOOD IS USED IN THE BODY

Those chips, sausages, beans and ice cream have now become quite different. They are now used by the body as amino acids, simple sugars, fatty acids and glycerol.

Amino acids are built up again into proteins and used for growth, body-building and repair. Extra protein may be used as extra energy reserves. Too much protein is dealt with by the liver, where the nitrogen part is removed and sent as waste to the kidneys (page 97).

Simple sugars are used for internal respiration in the cells to produce energy for all body functions. Extra sugars may be changed by the liver into **glycogen** and stored around the muscles and the liver. Too much sugar is changed to fat and stored in fat cells.

Fatty acids and glycerol are used for tissue growth and the healthy working of the cells. Extra fats may be used in tissue respiration to produce energy. Too much fat is stored in the fat cells of adipose tissue.

Between the large and small intestine is another sphincter. It opens to let roughage and undigested foods into the large intestine. The **appendix** is near this sphincter. It's a small worm-like structure which hasn't much use. It may become infected, swell up and have to be removed. The operation to remove an infected appendix isn't serious.

The large intestine (colon, rectum, anus)

The large intestine is not long, about 90 centimetres, but it is gathered up like a concertina. It is called large because it is so wide. There are no digestive glands in it. Its function is to deal with the roughage and undigested food, bacteria and dead cell linings, mucus and the water that are all that is left of the delicious meal. A great deal of water has been used to break down the foods and it must be returned to the body. So the blood vessels in the wall of the **colon** re-absorb this water and any useful salts into the

CHART OF FOOD IN THE BODY

Proteins	Carbohydrates	Fats
Animal proteins Plant proteins	Starches Sugars	Animal fats Plant oils
Digested in:		
Stomach Small intestine	Mouth Small intestine	Small intestine
Digested by:		
Pepsin from stomach glands Rennin clots milk Enzymes from pancreas Enzymes from small intestine	Ptyalin from salivary glands Enzymes from pancreas Enzymes from small intestine	Enzymes from pancreas. Enzymes from small intestine. (Bile from liver emulsifies fats.)
Digested into:		
Amino acids	Simple sugars, glucose.	Fats and fatty acids and glycerol.
Taken into:		
Blood capillaries in villi	Blood capillaries in villi	Lacteals in villi. (some fats pass into blood capillaries.)
Used as:		
Proteins for growth, body building, replacing old cells	Energy foods for cell respiration.	Energy foods for healthy tissue function.
Stored as:		
Very little is stored but is converted to energy food	Glycogen for a while in liver and muscles	Fat cells in adipose tissue
Too much is:		
Changed into urea and passed out of the body as waste.	Changed to fat cells and stored in adipose tissue	Stored or used as extra energy food.

bloodstream. This makes the waste matter, **faeces**, more solid and dry.

Peristalsis in the colon is slow and it isn't until the faeces pass into the **rectum** that we feel the need to go to the toilet. The **anus** is the last sphincter of the food canal and it opens to let out the faeces.

GETTING RID OF WASTE FOOD (DEFAECATION)

Babies and small children have no control over the muscles of the rectum. It takes 2 years or more for them to learn how to control them. While they are learning, they should not be punished for 'being dirty' or 'having an accident'. Sometimes very old people may lose control over these muscles. This is very upsetting for them.

Each person's body works at a different rate, so there can be no rule about how often we should go to the toilet. Some people go two or three times a day, others once a day, some once every 3 days. It used to be thought that if you didn't go to the toilet each day the poisons in the faeces would get back into your body and poison you. This is not true.

CONSTIPATION (PACKED TOGETHER)

It is better if you do have regular toilet habits or you may get constipated. This happens when the faeces have stayed too long in the colon and become hard and dry. They are then difficult to pass out of the body. **Laxatives** work by irritating the lining of the colon or by softening the faeces. It is not wise to use laxatives often or the muscles of the rectum lose their tone. Without muscle tone you get more constipated and you end up having to depend on laxatives. If you suffer from constipation:

a. Go to the lavatory when you have plenty of time.
b. Eat more foods with roughage in them (page 77).
c. Get more exercise for good muscle tone.
d. Drink lots more water and fluids.

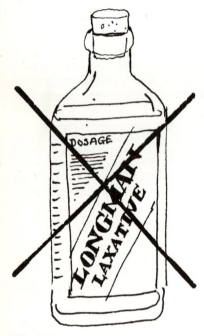

Fig 7.9

DIARRHOEA (FLOWING THROUGH)

This is often called the 'runs' as the faeces are liquid or semi-liquid. Diarrhoea can be caused by different illnesses. It is also caused by eating bad food. The germs multiply in the intestines where they are attacked by the white blood cells. This causes a lot of irritation of the colon. The peristaltic waves sweep the waste matter along too quickly and there is not enough time for the water to be absorbed back. People with diarrhoea must drink plenty of liquids to put back all the water they are losing. Important salts may also be lost. Diarrhoea stops when the infection clears up. It is a dangerous disease in babies and young children as they quickly become dehydrated and ill. A small child who has diarrhoea for more than 1 day needs medical attention.

Health and hygiene of the food canal

Food is one of the main ways in which germs get into the body.

1. Buy fresh foods.
2. Wash uncooked food thoroughly.
3. Meat and fish must be cooked right through.
4. Wash hands before preparing a meal.
5. Wash hands after going to the lavatory.
6. Wash hands before a meal.
7. Go to the lavatory regularly.
8. Chew food slowly and well.
9. Eating in a hurry causes indigestion.
10. Drink plenty of fluid with meals.
11. Don't row at mealtimes, enjoy them.
12. Unhappy feelings cause indigestion.
13. Don't treat your stomach like a dustbin.
14. Slimmers should eat less but have a balanced diet.

The liver

The liver is large, reddish brown and lies across the top of the abdomen. It is rich in blood as it has its own supply from the heart and all the blood from the hepatic portal vein from the small intestine. It is an extremely important organ and has many many different functions. Some of them are:

1. It produces bile which emulsifies fats.
2. It helps control the amount of sugar in the blood.
3. It filters the blood and controls the amount of blood in the body.
4. It stores iron.
5. It helps to make proteins for blood clotting.
6. It stores the fat-soluble vitamins, A, D and K.
7. It makes poisons in food, drugs or alcohol harmless.
8. It produces a lot of body heat. While we are sleeping, the heat made by the liver keeps us warm.
9. It changes excess amino acids into urea which it sends to the kidneys.

These are only a few of the functions of the liver. It is important you learn them.

Questions and things to do

You will need to study a dissected rat, rabbit or any small mammal. As you do so, make sure you understand each part of the food canal you are examining. Make notes and drawings of everything you see. How are all the coils of the small intestine kept in place? Find the blood capillaries leading to the hepatic portal vein. Find the digestive glands.

Pancreas is eaten as 'sweetbread', intestines are eaten as 'tripe'. Have you ever tasted them? If not, you might enjoy cooking and eating them. What nutrients do you think they would contain? Find out what 'haggis' is.

Test the digestive action of saliva on starch
Collect five test-tubes and label them A, B, C, D, E. Rinse
out your mouth with water to get rid of any waste food.
Then collect your own saliva in two clean test-tubes. Boil
the saliva in one of the tubes.
Put fresh starch solution in test-tubes A, B, C, D, E.
Add the unboiled saliva to test-tubes A, B and E.
Add the boiled saliva to test-tubes C and D.
Place the test-tubes A, B, C and D in a water bath at 37 °C
for at least 5 minutes.
Place the test-tube E in the refrigerator or any cold place.

Add iodine solution to test-tubes A and C and boil.
Add Benedict's solution to test-tubes B and D and boil.
Add iodine solution to test-tube E and boil.

If a blue colour does not appear in the tubes tested with
iodine, what do you think has happened to the starch
solution?
If a red colour appears in the tubes tested with Benedict's
solution, what do you think has happened.
Give reasons for the results you got from using the boiled
saliva.
What has happened to test-tube E? Why do you think this
is?

1. Starting with the mouth, list the correct order of the
   organs and glands of the food canal.
2. Explain clearly how food is moved along the food canal.
3. Draw and write about the position, structure and
   functions of the stomach.
4. Copy out and learn the facts about the liver. Why is
   animal liver such an excellent food?
5. Write a little about each of the following: sphincters,
   enzymes, secretions, ducts, lacteals.
6. How is digested food removed from the small intestine?
7. What are the functions of the colon?
8. You have eaten a meal of beef, potatoes and fresh salad.
   Write down where these different foods are broken down
   and by what enzymes. Then explain clearly how they are
   used by the body. This question may take a long time to
   answer.
9. Copy out and learn the chart of food in the body (page 93).
10. Draw in the gullet and stomach, in their correct positions,
    on one of the skaters.

Chapter Eight

Water in the body

Whether we are awake or asleep, sitting quietly or rushing about, the systems of our body go on working. Chemical changes keep happening. Cells, tissues and fluids are always being mended or replaced as they wear out and we make new ones. Living protoplasm is changing day by day and many of these changes produce waste. Some of these waste products are poisonous or **toxic** and must be got rid of by the body. Getting rid of waste products is called **excretion**.

We know the lungs get rid of, excrete, carbon dioxide – the waste product of internal respiration. The skin also helps to excrete small amounts of waste products. Other waste products are excreted by the **urinary system**. Nitrogenous waste from protein is one of the main things excreted. Protein is made up of different complicated chains of molecules which take a lot of breaking down. Nitrogenous waste from protein is changed in the liver into **urea**. Urea is then sent into the bloodstream and down to the kidneys. The liver also works on toxic things such as alcohol and drugs (page 95) before they are sent to the kidneys. The **kidneys** are able to remove all these things from the bloodstream by filtering them out of the blood.

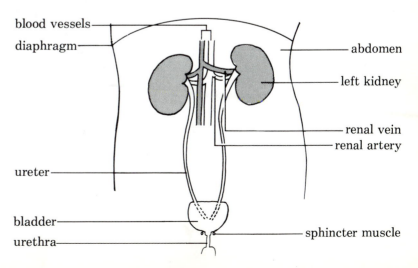

Fig 8.1 The urinary system

blood vessels
diaphragm
abdomen
left kidney
renal vein
renal artery
ureter
bladder
sphincter muscle
urethra

THE URINARY SYSTEM

Notice the position of the two kidneys high up at the back of the abdomen, one on each side of the spine. Boxers are not allowed to punch their opponents in this area, 'the small of the back', as a blow on the kidneys will cause serious damage. The kidneys are bean-shaped, about 10 centimetres long, and dark red because of their rich blood supply. Find the **renal artery** and the **renal vein** which bring blood to and take it away from the kidneys. The two **ureters** lead down into the **bladder** and there is a short tube, the **urethra**, from the bladder to the outside.

The kidneys

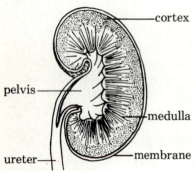

Fig 8.2 The kidney

Get a kidney from the butcher. Feel the tough protective membrane around it. Cut the kidney in half lengthwise. Use a hand lens to study the three layers. Find the outer dark part, the **cortex**; the middle lighter part, the **medulla**; and the inner part where the ureter begins, the **pelvis**.

STRUCTURE
The nephrons (tubes)
The cortex is made up of millions of tiny 'cups', only one layer of cells thick. The cups are called **Bowman's capsules**. Each capsule has a long **nephron** or tube leading from it. This nephron dips up and down through the cortex and the medulla. It then joins up with all the other nephrons in the pelvis.

The blood vessels
The renal artery enters the kidney and divides into very small blood capillaries. Each capillary forms a tiny cluster, called a **glomerulus**, inside a Bowman's capsule. It then makes a network of tiny blood vessels around the loops of the nephron. Finally, all the capillaries join up together to leave the kidney in the renal vein.

HOW THE KIDNEY WORKS
The blood enters the kidney by the renal artery. This divides into tiny capillaries which enter the Bowman's capsules. Most of the liquid part of the blood is forced out of the capillaries and in through the thin walls of the capsule. About 100 litres of watery fluid are forced into the capsules each day. Most of this fluid must be taken back into the bloodstream.

As it passes along the loops of the nephron, the network of blood capillaries takes back or **re-absorbs** the fluid. It leaves the waste products and a small amount of water in the nephrons. The network of blood capillaries having filtered out the waste and re-absorbed its fluid, joins up to form the renal vein. The filtered or 'cleansed' blood continues on its travels around the body (page 43).

What is left in the nephrons is emptied into the pelvis as **urine**.

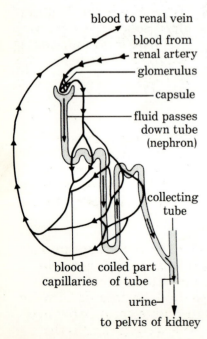

Fig 8.3 How the kidney works

The ureters

The two ureters leave the pelvis of each kidney and enter the **bladder** at its lower end. The ureters are about 25 centimetres long and urine is moved down them by peristaltic waves. Urine enters the bladder in tiny jets, but we don't feel this happening.

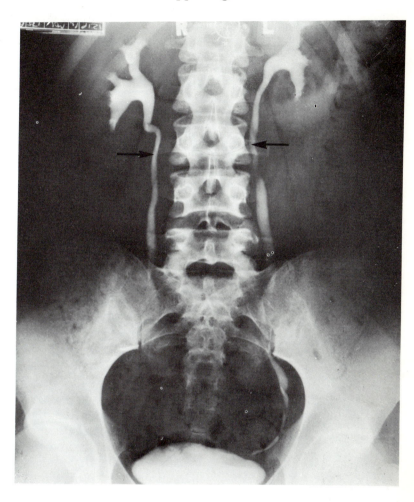

Fig 8.4 X-ray showing the ureters (arrowed) leading from the kidneys to the bladder

The bladder

The bladder is a hollow muscular bag which can stretch to hold 300 cubic centimetres or more of urine. It is a reservoir in which the urine slowly collects. There is an inner and an outer sphincter to hold the urine in. When the bladder is nearly full we feel the urge to pass water, to **urinate**.

The urethra

This is the urine tube from the bladder to the outside. You must be careful not to confuse it with the ureters as the words are so alike. The urethra is quite short in the female, between 3 and 4 centimetres long. It leads from the bladder to the first opening of the **vulva** between her legs (page 169). The urethra in the male is longer as it is the tube leading down the middle of the **penis**.

Passing urine (urinating)

When the bladder is nearly full, messages are sent to the brain and spinal cord. The control on the sphincters is cancelled and they relax. The diaphragm is forced down and the abdomen walls contract. The muscles of the bladder also contract and the urine is forced down the urethra. It is more complicated than people think, as nerves, muscles and sphincters have to work together at the right time. Babies have no control over all this. They can learn to sit, stand and walk before they can learn to control their bladders.

This is important to understand so that, when you are a parent, you will not punish a child or make it feel ashamed if it 'wets itself'. It is stupid and cruel to blame a child for doing something it has no control over. It is like telling-off a blind man because he cannot see. Those children, whose parents have been stupid about toilet training, may develop nervous problems as they get older.

Water balance in the body

Urine is made up of about 96 per cent water, 2 per cent urea and 2 per cent other salts. The urea is the waste product from protein digestion. The other things include excess salts, harmful substances and anything which upsets the delicate balance of the fluids of the body. Sometimes the urine will have more water in it and sometimes it will have less. This is because the kidneys help to control the water balance in the body. How much water do we lose each day? Go back to page 74 to check your answer. Under normal conditions, we pass about 1 to $1\frac{1}{2}$ litres of urine a day. But if we sweat a lot because we are in a hot climate or are doing heavy exercise, and if we don't drink more fluids to make up the water lost in sweating, then the kidneys will keep back more water and the urine will be stronger. But if we drink too much liquid, especially tea, coffee and cocoa which cause an increase in urine, then the kidneys will pass a lot of water through. The urine will be paler in colour as it has so much more water. Certain diseases can cause a change in the amount of urine, as can nervousness and excitement.

Health and hygiene of the urinary system

The kidneys are very important organs. We can live with only one working, but if both kidneys failed we would die within 2 weeks. People with damaged kidneys have their blood filtered, the fluids regulated and the water balance controlled by an artificial kidney, a **dialysis machine**. Unfortunately, these machines are very expensive to make and there are not always enough to go round. Raising money to buy a kidney machine for your local hospital may be something you can do to help.

A damaged kidney can be replaced sometimes by a healthy kidney from another person. This is not simple as

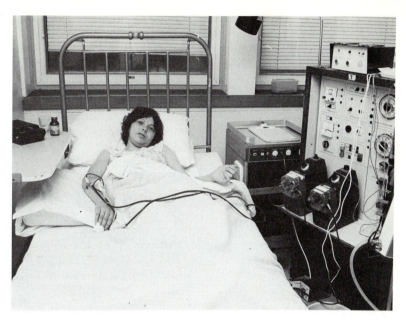

Fig 8.5 A patient on a dialysis machine

blood groups and tissue types must be as similar as possible. Not everyone with a damaged kidney has an identical twin! If you are interested in learning about **kidney transplants** ask your local health education office to give you details of the kidney donor scheme.

Certain diseases can be found by testing a patient's urine. The kidneys filter out harmful substances and they are passed out in the urine. When urine is tested, **analysed**, the doctor can work out what is wrong in the body from what is found in the urine. As the kidneys also regulate the delicate balance of fluids, an excess of anything will show up. A lot of sugar in the urine may mean the patient has **diabetes** (page 163). Alcohol is also found in the urine which is why suspected drunken drivers may be asked to give a specimen of urine at the police station.

CLEANLINESS

Urine is usually free of germs but faeces are not. As the anus is so close to the external excretory organs, the whole area should be washed carefully every day. Girls need to be especially clean to prevent any germs getting to the tiny opening of the urethra and travelling up to the bladder. Germs will breed in the bladder causing **cystitis**. This gives burning pain when passing urine and the feeling of wanting to pass urine all the time.

Other infections of the urinary system may have the same symptoms for boys and girls. *Always* go to your doctor if you have any trouble or pain in passing urine. It is much easier to cure a slight infection than one which has lasted for ages and may have caused a lot of trouble.

Excretion is the method used by the body to get rid of waste products of metabolism and excess or harmful things.

These methods are:

a. The lungs excrete carbon dioxide and water.
b. The kidneys excrete urea, salts, excess and harmful substances as urine.
c. The skin excretes water with very small amounts of salts and urea as sweat.

Faeces, got rid of from the colon, are not really excretions. They are made up of food we have not been able to digest; that we have not been able to absorb into our bodies. (They do have dead cells from the walls of the intestines and bacteria from the colon, but as the main bulk of faeces is unwanted food they are not included in excretions.)

Questions and things to do

1. Draw and label the urinary system.
2. What are the functions of the kidney?
3. Using diagrams, explain how the kidney works.
4. Collect a kidney donor card from your doctor or health education officer. Write a few lines explaining its use.
5. Explain clearly the difference between urea and urine.
6. Explain clearly the difference between a ureter and the urethra.
7. Make a list of the waste products of metabolism. Beside each, write down where it is made and how it is excreted.
8. Copy out and learn the meaning of 'excretion'.
9. Why may a doctor ask for a specimen of urine?
10. Why is it important to be gentle and patient with a small child when it 'wets itself'.
11. Draw the kidneys in their correct place on one of the skaters.

Chapter Nine

Skin and the control of body temperature

We usually only think of skin as something which makes us more attractive when it is healthy, or less attractive when it is spotty or greasy or dry. But there is far more to skin than this. It is an amazing organ, spreading over the whole of our body and doing a great number of things.

Fig 9.1 The skin

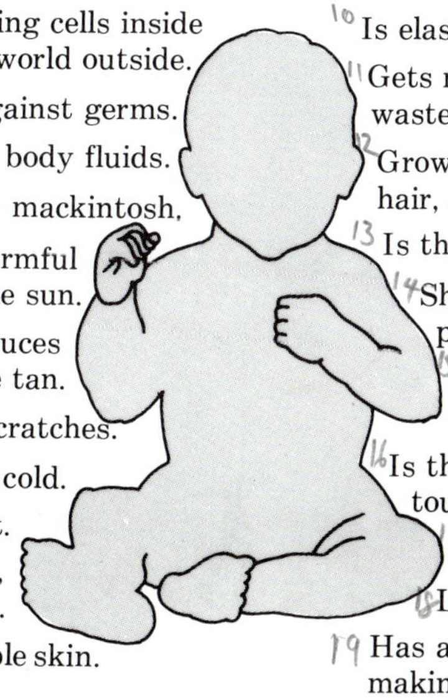

1 Protects soft living cells inside from the harsh world outside.

2 Acts as a barrier against germs. Keeps in precious body fluids.

3 Keeps out the wet, like a mackintosh.

4 Shields us from the harmful rays of the sun.

5 In strong sunlight, it produces a protective tan.

6 It heals wounds, small cuts or scratches.

7 Keeps in heat when we are cold.

8 Lets out heat when we are hot.

9 It helps to make Vitamin D, in sunlight.

Makes oil for shiny hair and supple skin.

10 Is elastic, springing back into shape.

11 Gets rid of small amounts of waste products.

12 Grows body, face and head hair, and nails.

13 Is the main organ of physical attraction.

14 Shows our feelings, darker with anger, paler with fear.

15 Shows our health, greyer when ill, bluer when short of oxygen.

16 Is the main organ of physical feelings, touch, pain, pressure etc.

17 Lasts well all through life.

18 Loses elasticity in middle age.

19 Has a special system for making and shedding cells.

The structure of the skin

The skin has two main layers, the outer **epidermis** and the inner **dermis**.

THE EPIDERMIS

Wet your finger and rub it briskly up and down your arm. You will see tiny flakes or rolls of dead skin coming away. The whole surface of our body is made up of these dead cells. It is weird to think the skin we stroke, cuddle or kiss is quite dead. But it is true. Living cells would be destroyed if they came into contact with the air around us. To prevent this, the epidermis has a very clever system of making its own cells and then killing them off before they reach the surface of the skin.

Study Figure 9.2 and find the **germinating layer**. The function of this layer is to make new epidermis cells all the time. These cells get pushed slowly towards the surface to make room for the next cells being produced. As they travel outwards, the cells become thinner, flatter and full of **keratin**. Keratin is a hard substance used to make hair and nails. By the time the cells reach the surface, they are scaly and quite dead. They stay on the surface of the skin for a time, doing their work of protecting the delicate inner cells. Then they wear away and fall off. Millions of dead cells are shed from the surface of our skin each minute. So, the cells of the epidermis are always being made, moving outwards, becoming filled with keratin, dying and being worn away.

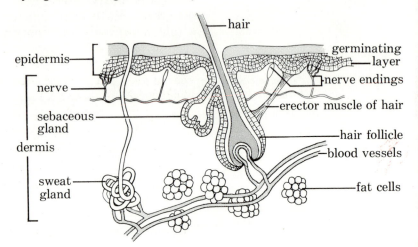

Fig 9.2 A section through the skin

Feel the skin on the palm of your hand. Like the skin on the soles of your feet, it is tough. This is non-hairy skin and has a thick epidermis. It is also lined and has a pattern of ridges at the tips which form the fingerprints (Figure 9.3).

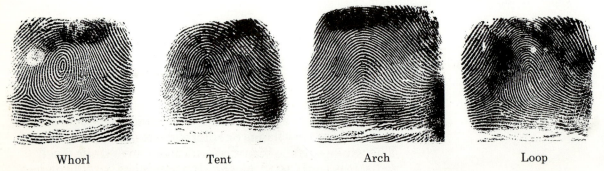

| Whorl | Tent | Arch | Loop |

Fig 9.3 Fingerprints

Make a copy of your fingerprints and compare them with those of the rest of the class. No two are alike. Each person in the world has a different pattern of prints. Palm prints, sole prints, finger and toe prints are formed about 5 months before we are born!

The skin on the rest of the body is hairy skin. It has a thinner epidermis and contains the pigment melanin which determines our skin colour. There are no blood vessels in the epidermis so food and oxygen are passed in tissue fluid from the dermis.

THE DERMIS

This is a deeper layer of inner tissue. Study Figure 9.2 carefully. Notice the blood vessels, nerve endings, sweat glands, sebaceous glands, hair follicle and erector muscle of hair. Just below the dermis are the fat cells, the adipose tissue (page 78).

Temperature control of the skin

Temperature contol is an extremely important function of the skin. The best temperature for our bodies to work at is between 36.8 and 37°C. The cells, tissues, fluids and enzymes need this amount of warmth to function properly. Too much or too little warmth will upset the metabolism of the body. Extremes of temperature inside the body will lead to death. The skin helps control the amount of warmth in the body and keeps our temperature at a steady level. Having a *constant body temperature* means we can live in very cold or very hot countries.

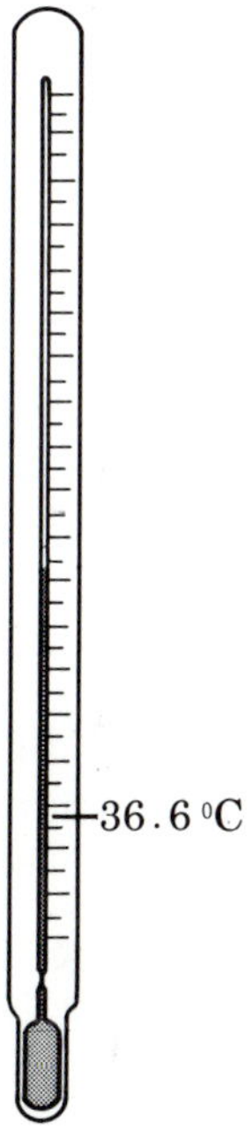

LEARN HOW TO USE A CLINICAL THERMOMETER
Read it. Always hold the end opposite the bulb.
Shake down the mercury if it shows a temperature reading.
Wipe the bulb end with clean cotton-wool.
Place the bulb end under the tongue.
Leave in closed mouth for 3 minutes.
Remove thermometer and read the temperature.
Shake down the mercury.
Wipe the bulb end clean before replacing in the case.
(At first, you may find it difficult to read the thermometer. Turn it very slowly till you can see the dark bar of mercury.)

There is no such thing as a 'normal' body temperature. It varies between 36°C and 37°C. Our temperatures are lowest in the mornings and rise slightly during the day. Babies and small children have higher temperatures than adults. Old people have lower ones.

A person's temperature should not be taken after a hot or cold drink, a bath, or a lot of exercise. Why do you think this is? It is important to understand that a high temperature is a *symptom* of an illness, not an illness itself. You only need to take someone's temperature when there are other symptoms of illness.

We make our own heat, we use it to keep functioning at the right temperature and we get rid of what we don't

Fig 9.4 A clinical thermometer

36.6 °C

want. The balance of heat in our bodies is controlled by our brain. If the temperature of the blood rises or drops below our normal level, the brain sends messages to the skin and other parts of the body to put it right.

Body heat comes from:	Body heat is lost by:
Our metabolism; the chemical changes which happen in the body. The liver produces a lot of heat as it has so many different functions. Other chemical changes such as cell respiration, digestion and filtration in the kidneys produce heat. Contracting muscles, both voluntary and involuntary, produce heat. External heat — fires, sunshine, central heating.	Blood takes the heat from the cells to the skin and heat is passed into the air. Movement—especially sport and hard physical work. The air we breathe out has been warmed to body temperature. Urine and faeces are excreted at body heat. Sweating.

TOO MUCH HEAT

1. The tiny blood vessels in the dermis **dilate**, that is they get wider. More blood can pass into them so more heat from the blood can be passed to the surface of the skin to cool us down. This is why fair-skinned people go red in the face when they are very hot.
2. The sweat glands start working. Fluid from the blood is passed into the sweat glands and travels up the ducts to the surface of the skin. As the sweat evaporates we become cooler. Beads of sweat can be seen on the forehead, the neck and face.

TOO LITTLE HEAT

1. The tiny blood vessels in the dermis contract, less blood passes through so less heat is lost. In extreme cold, the blood vessels close completely, so no heat is lost. The skin goes white and dead-looking.
2. Messages are sent to the muscles to begin shivering. These tiny contractions of muscles produce heat energy, helping to raise the body temperature.
3. The erector muscles attached to the hair of the skin contract. The hair is pulled erect and we get 'goose pimples'. This is not much use in keeping us warm as we have such fine body hair. But furry animals and birds can fluff themselves up and trap warm air this way.
4. The liver makes more heat and the sweat glands stop working.

Problems of overheating

Overheating can be caused by many things – a hot humid climate, disease, over-exercise, and so on. It can also be caused by a coat of gold paint! Goldfinger's lady friend died because (*a*) she was not able to lose heat from the surface of her skin and (*b*) she was not able to lose heat by sweating. Her inner temperature rose too high and killed her.

HOW WE LOSE HEAT BY SWEATING

Dip your hand in warm water and then wave it around to dry. You will feel the skin go cool and then quite cold. When water **evaporates**, it changes from a liquid to a gas – from water to **water vapour**. Heat energy is needed to evaporate water and the heat has been taken from your hand. Sweating cools us because, as sweat evaporates, it takes heat from our skin. In hot climates, or on hot days in cool climates, sweating is the most important way of losing heat. If the air around us is hot, we can't pass out heat into it, so we lose heat by evaporation of sweat.

HUMIDITY

At times the air is full of water vapour. During the rainy season, it is difficult for people in tropical climates to keep cool. As the air is soaked with water vapour, their sweat will not evaporate. If they do hard physical work, they may suffer from **heatstroke**, unless they rest and cool off as soon as they get hot.

At a school assembly, when the windows and doors are closed, students may get very hot, feel giddy or even faint. This happens because the air in the hall becomes soaked with water vapour from the people sweating *and* breathing out water vapour. Once the air in the hall reaches total humidity, no more sweat can be evaporated. Everyone begins to feel very uncomfortable as their cooling system cannot work. Whenever a lot of people are gathered together in a room, windows and doors must be opened wide to allow humid air to be removed and fresh, dry air to be circulated.

TOO MUCH SWEATING

When we sweat, we lose mineral salts, the most important one being sodium chloride, which is common salt. If we sweat a lot, we must put back the salt we have lost, or we get cramp in our muscles. Miners and stokers and others working in very hot conditions take salt tablets or eat extra salt with their meals.

Problems of overcooling

Overcooling happens when we lose more heat than we produce. It is a serious problem in babies and elderly people living in cold climates.

COLD BABY SYNDROME

Babies have a large surface area for their body size so they lose heat very easily. Because their brain cells are not fully developed, the temperature-controlling part of their brain is not yet working properly. The skin and blood

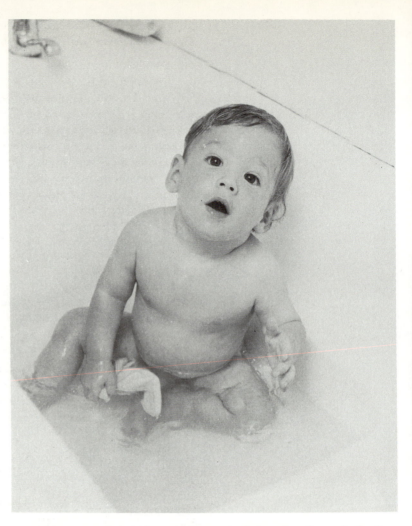

Fig 9.5 The room has been well heated before the baby's bath

vessels may not get the messages to let out or keep in body heat. Parents should know this as it is an important part of their job to keep their baby comfortably warm.

If a baby's body heat drops, he is not able to raise it as we can raise ours. A cold baby stays cold and slowly gets colder, becoming dangerously ill. A really cold baby lies quite still, refuses to eat, the hands and feet swell, and though the face looks pink the skin is cold to touch. He must be taken immediately to hospital so he can be gently, safely and slowly warmed again.

Cold Baby Syndrome can be prevented by making sure the baby's room is really warm. A temperature of 32 to 33°C is best for new-born babies. This is much higher than is usual in most homes, and the baby will only need a light blanket to cover him. Remember, extra clothes and blankets are *not warm in themselves*. They are useful to keep in body heat already there. They cannot warm a cold baby. But they will help a warm baby to keep in his warmth. The temperature in a baby's room should not

drop below 20 to 21°C, when he is warm and well
wrapped up. Check the temperature in your classroom.

HYPOTHERMIA

This word, meaning below body heat, is used when elderly
people suffer from overcooling. It most often happens to
poor people who live alone. They have not enough money
to heat their home nor to eat a proper diet and they are not
strong enough to take exercise. These things, together with
their slower metabolism, cause the body temperature to
drop to a dangerous level. Their minds become confused as
they get tired and weak. It is sad to think that many old
people quietly freeze to death in cold weather.

Young active people hardly ever suffer from hypothermia
unless they are stranded in some cold place, such as a
mountain side. If possible, they should:

a. Find or make a shelter. Cold moving air quickly
    removes body heat.
b. Remove wet clothing. Damp clothes use body heat to
    dry out.
c. Put on *layers* of dry clothes. Air warmed by body heat
    is trapped in the layers.
d. Keep moving. Muscle contraction produces body heat.
e. Keep awake. The body's metabolism, which produces
    body heat, drops when we sleep.

FROSTBITE

Our **extremities** – fingers, toes, lips, ears and nose – are
most likely to suffer from frostbite. The tiny blood vessels
contract away from the skin. Food and oxygen cannot get
through to the cells and waste is not removed. Slowly, the
tissues of the frost-bitten area die and begin to rot, to be
attacked by germs. In severe frostbite the person may have
to have the finger or toe cut off, **amputated**, to prevent
infection from the dead tissue getting into the
bloodstream. The blood vessels also get badly damaged so
that fluid leaks from them and blood clots are formed.

**Helping control body
temperature**

It is important to remember that whether we *feel* hot or
cold, our body temperature remains constant. (Except in
the case of severe overheating, cooling and disease.) Our
skin may feel too hot or too cold, but our inner temperature
stays the same. Nerve endings in the skin send messages to
the temperature-controlling part of the brain. The brain
sends out instructions to warm us up or cool us down.

Man has found ways of helping to control body
temperature (page 110).

**More facts about
the skin**

SKIN COLOUR

Our skin colour depends on the amount of **melanin** in the
skin. Melanin is a dark pigment which protects the skin

a

b

c

d

Fig. 9.6 a Loose cotton clothes absorb sweat and let out body heat

b Layers of wool or fur clothes keep in body heat

c These people are protected from the fierce direct heat of the sun and cooled by the breezes which can blow through the building

d The walls and roof of this house are being lined with insulating material to stop the heat escaping into the cold air

from the ultra-violet rays of the sun. We all have pigment producing cells and we all tan in sunshine. Dark people get darker, fair people take longer to make the melanin before they tan. They should not stay in strong sunlight for long periods until the melanin has begun to be produced.

VITAMIN D
The body can make its own vitamin D in sunlight.

ADIPOSE TISSUE
The fat cells stored under the skin help to prevent the body from losing heat. People in very cold climates eat a lot of animal oils, which are stored in the adipose tissue, to keep them warm. Fat people in hot climates may have more trouble in getting rid of extra body heat than thinner people.

SENSE ORGAN
Find the nerve endings in Figure 9.2. They are all over the surface of the body, many more in places such as the lips and fingertips and less in places such as the middle of the back. Nerve endings receive messages of temperature, pain, pressure and touch. The messages are sent to the brain or spinal cord and are dealt with. The skin is a very important sense organ as it keeps the brain informed about what is happening on the surface of the body (page 134).

THE HAIR
There are three main types of hair: straight, wavy and tightly curled.

Fig 9.7 The three main types of hair

The average head has about 100 000 hairs and between 30 and 40 drop out each day. Most of the skin is covered with hair though it may be so fine we can hardly see it. Hair, like nails, is made from the germinating layer of the epidermis. It is packed with keratin (page 104) and is quite dead. Each hair is shed from its follicle after a few years and a new one grows in its place. In some men, the hair follicles stop making new hair and they become bald. In old age some women may also lose their hair. There is no cure for baldness as yet, though men spend a lot of money on creams, lotions and massages to try to stop it. Eyebrows and eyelashes protect the eyes from rain, sweat and dust particles. Otherwise, hair is not very useful except to make us look more attractive.

THE SEBACEOUS GLANDS
They make an oily fluid called **sebum**. Sebum is secreted into each hair follicle and travels to the surface of the skin. Its function is to keep the hair and skin oiled and smooth.

THE SWEAT GLANDS
These glands help the body get rid of heat. After childhood more glands begin to work in the armpits, the groin, the palms of the hands and the soles of the feet. The sweat they produce is heavier, gets an unpleasant smell quickly and is often made when our emotions are upset, e.g. when we are shy or embarrassed, angry or afraid. Once the water has evaporated from sweat, a greasy deposit of urea and mineral salts is left on the skin.

Health and hygiene of the skin

During **adolescence**, the 'teen' years, a great many changes take place in the body. The glands may become over-active, the skin may be too dry or too greasy, and the hair lank or brittle. Most adolescents have to put up with at least one of these things happening. So care and cleanliness of the skin, which is always important, becomes even more so in adolescence.

1. *Always clean any break in the skin and dab on antiseptic lotion.* Our skin is like a suit of armour in that it defends the delicate tissues of the body from harm. It protects us from germs and dirt in the air around us. It can only do this when it is unbroken. Any cut, scratch or tear in the skin, however small, will let in the germs from outside.

2. *Cover wounds to prevent germs entering.* The bigger the wound the longer it takes to form a scab. During this time, it is wise to cover the wound to prevent any infection getting into it. With very deep cuts, the edges of the skin must be pulled up and sewn or clipped together.

3. *Burns are serious injuries to the skin.* Small burns, those no bigger than a bottle top, can be treated by holding the burned area under the cold tap for 10 minutes. Creams and lotions should not be used on burns. A larger or deeper burn which damages the germinating layer must be treated at the hospital. Burns may cause severe shock and terrible scarring.

4. *Keep the surface of the skin really clean.* Sebum, sweat and dead cells mix with the dust in the air. Germs breed

HEALTH AND HYGIENE OF THE SKIN

Body			
 Wash body all over once each day.	 Drying with a towel rubs off dead skin, helps blood circulation.	 Nails scrubbed clean daily and cut often.	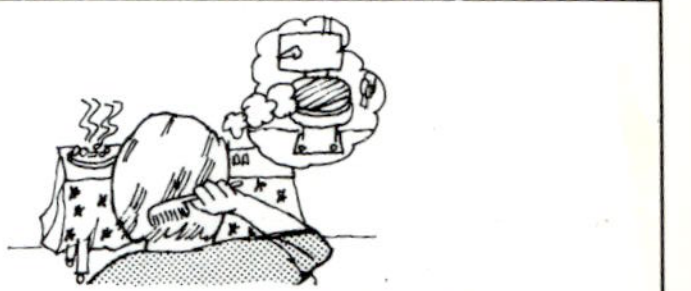 Wash hands *after* combing hair, going to the toilet and *before* cooking or eating.

Face			
 Wash often in warm soapy water.	 Rinse with cold water for two minutes.	 Gently pat spotty skin dry. Put cream on dry skin.	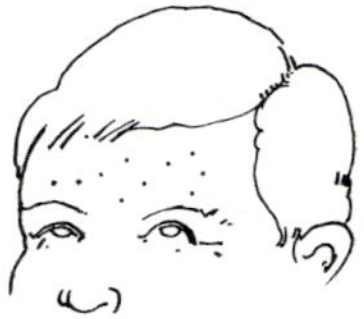 Squeezing spots can cause acne.

Hair			
 Wash greasy hair often as dust and dirt stick to it.	 Add a little oil to dry or brittle hair. Wash only once a week.	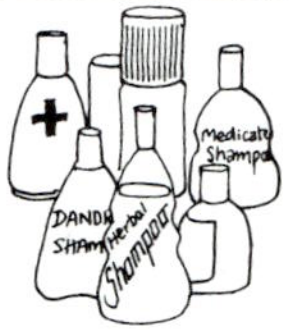 Dandruff can be treated with special shampoos. Dandruff may cause acne.	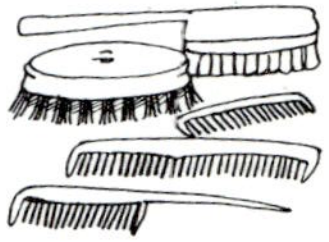 Don't borrow or lend brushes and combs. Wash them often

Clothes			
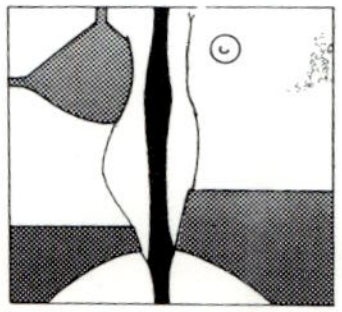 Change underwear every day	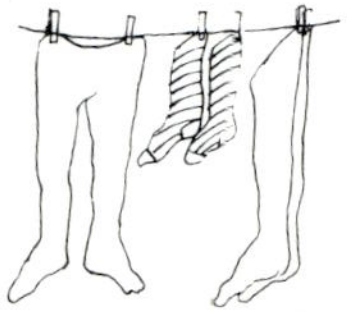 Wash socks, stockings or tights every night.	 Don't borrow or lend clothes and shoes.	 Outer clothes also get full of bacteria, dead cells, sebum and sweat. Clean often.

Fig 9.8 Health and hygiene of the skin

on this mixture and produce the smell we call **body odour**. Daily washing with soap and water removes waste and dirt from the surface of the skin. Rubbing briskly with a towel removes dead cells and tones up the blood circulation. Washing and drying the skin help to control body odour. A **deodorant** or **anti-perspirant** can be used under the arms and on the feet.

5. *Caring for adolescent skin.* The openings of the sebaceous glands may get blocked with dried sebum. This causes blackheads, whiteheads, spots and pimples on the greasy areas of the face, neck and shoulders. Spots and pimples should be left alone; blackheads and whiteheads can be removed if *the greatest care is taken.* Scrub the nails thoroughly first. Press around the area gently to squeeze out the solid sebum. Dab on an astringent, rubbing alcohol, to disinfect the area. Bathe in cold water. **Acne** is a nasty infection of the skin caused by the changes in adolescence, especially in oily skin. It is often made worse by careless picking and squeezing of spots. Sharp nails break open the skin, dirt under the nails gets into the skin and the germs in the sebum are carried on the nails from spot to spot so the whole face becomes infected. Acne can be treated by a doctor. Otherwise, wash the skin whenever possible, splash with cold water for 2 minutes, pat gently dry and *leave alone.*

6. *Parasites of the skin.* An unwashed skin makes a lovely home for tiny animals and fungi. A **parasite** lives off other living things. Fleas, lice, mites, ticks and fungi are parasites living off people. It is quite usual for children, at some stage, to catch one or more of the skin parasites. **Infestation** of the skin is common in industrialized areas where people live close together, as it passes easily from one person to another. We also catch skin parasites from rats and mice, from pets and from stray animals.

When we are infested, we feel hot, itchy and sore. We are also angry and upset because we feel 'unclean'. Many people don't go for treatment because they are ashamed. Any infestation *must* be treated immediately as there is a risk that some parasites may carry other more serious diseases. The other important reason for treatment is that parasites break open the surface of the skin. Germs quickly get in and start to breed. This is called a **secondary infection**, when one disease causes another one. Thorough daily washing cuts down the chance of catching skin parasites.

Some skin infestations

FLEAS

The flea feeds on blood so is usually found very close to the skin, in underclothes, nightwear, sheets, blankets, carpets

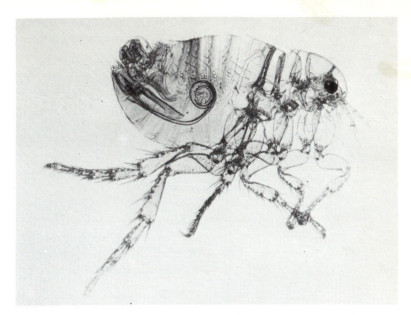

Fig 9.9 Photograph of a human flea

and furniture. It has very powerful legs and can easily jump from one person to another. Like other insects, it has horny tubes instead of a mouth. It pushes the points of the tubes through the epidermis and into the dermis. It sucks up blood through one tube. At the same time it pours saliva down the other tube. The saliva has a special substance which stops the blood from clotting while the insect feeds. If the flea is carrying a disease in its body or on its mouth-parts, the germs of the disease are injected straight into the person's bloodstream. **Rat fleas** and **human fleas** used to carry the disease called plague or black death.

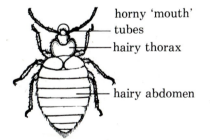

horny 'mouth' tubes

hairy thorax

hairy abdomen

Fig 9.10 Diagram of a bedbug

BEDBUGS
These are also blood-suckers and live mainly in mattresses They crawl onto the person at night to feed. They can live for a long time without food, hiding away under wallpaper and in tiny cracks in the walls. They do not usually carry diseases but their bites are very irritating. Scratching breaks open the skin which may cause secondary infection.

LICE
Lice actually live on the human body the whole time and have curved claws to give them a firm grip on the skin and hair. The **head louse** is very common among children, though many adults are also infested with them. The female lays about 60 eggs, or **nits**, each month at the base of the hair, especially round the back of the ears and neck. The female glues each nit to the hair with a strong cement. The nit hatches out in a week and is fully grown in 3 weeks.

    Body lice are not much different from head lice and may breed with them. **Pubic lice** or **'crabs'** live in the hair around the groin. In some countries lice carry a typhus disease and relapsing fever.

MITES AND TICKS

These are tiny blood-sucking creatures, not insects, which are responsible for carrying many serious diseases from person to person. The **itch mite** causes **scabies**, which is a very common infestation of the skin. The female burrows under the thin skin of the wrists, between the fingers and toes, the sides of the feet and under the breasts in women. She lays her eggs at the end of the 'tunnel' which can be seen from the surface as a thin, bumpy black line. The eggs hatch in a few days, crawl out onto the skin and begin to feed. They mate with the male mites which live on the surface of the body. Scabies cause great itchiness and it is difficult for the person not to scratch. Scratching helps to spread the mites and breaks open the skin.

FUNGUS PARASITES

There are several types of fungus which live on the skin. One type is called 'athlete's foot' and is common in boys whether they are athletes or not. The fungus is picked up

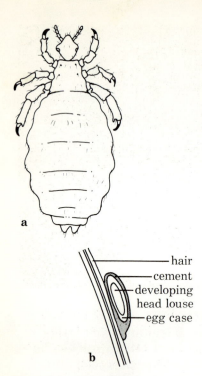

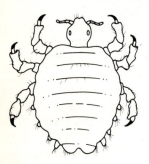

Fig 9.11   **a**   A head louse
           b   A nit glued to a hair

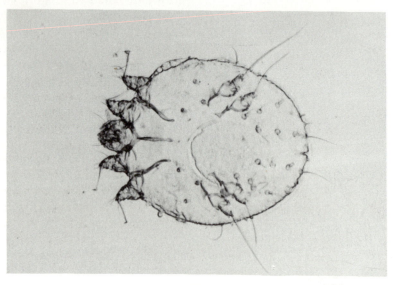

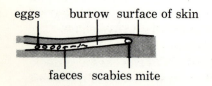

Fig 9.12   Diagram of a pubic louse

Fig 9.13   A scabies mite

Fig 9.14   Diagram of a scabies mite burrowing under the skin

from dead skin on towels and clothes, and from the dust in changing rooms and public swimming baths. It grows between the toes and can easily infect large areas of the foot.

Other types of fungal infection are called 'ringworm' and can be seen on the head, the nails, arms, legs and groin. There are no worms in ringworm. However, the fungus does often grow in a circle or ring. Ringworm can look horrid but it is not a serious disease and can be easily treated.

TREATMENT OF SKIN INFESTATIONS

As body parasites are passed by *contact*, the infested person

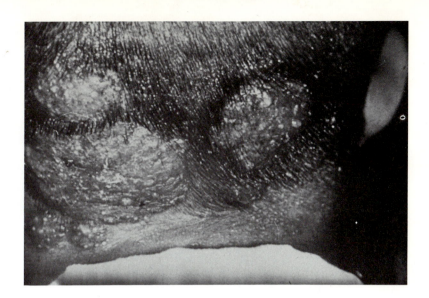

Fig 9.15 A nasty infection of ringworm of the head

must take great care not to infect others while he or she is being treated. Towels, clothing and bedding must be kept separate. Small children are kept away from school as they cannot be expected to understand about spreading infection. Older children and adults can go on working but must be careful not to spread the infection. Bites can be dabbed with antiseptic and sore, broken skin covered with a dressing.

DDT is an **insecticide** which kills most insects. It can be sprayed over bedding, furniture, walls, ceilings and floors. It is not much use just killing off the parasites. The eggs must be found and destroyed too. Spraying makes sure the insecticide gets into the dark corners and tiny cracks where the eggs are. Clothes can be sprayed with weak DDT, then washed in disinfectant and ironed with a very hot iron. The family pet can be dusted with a weak DDT powder as it may well be giving a home to fleas and other parasites. A few insects are becoming **resistant** to DDT and other insecticides may be needed. If a whole family suffers from infestation, the health authorities will de-infest the home and give advice and help against more attacks.

1. Fleas, bedbugs, ticks and mites can be killed by using DDT or a new insecticide.
2. Scabies are killed by painting the infected area with a special liquid. The liquid is left to dry and not removed for 2 days, to make sure the eggs under the skin are destroyed. Clothing worn during this time must be boiled and ironed, then thoroughly sprayed with DDT.
3. Ringworm can be destroyed in two ways: by spreading fungicidal cream over the infected skin or by swallowing tablets which attack the fungus from the bloodstream. During the treatment, clothes should be

Fig 9.16 Using an insecticide spray

boiled after each wearing, as the fungus can grow on material. Cats and dogs may carry the fungus which causes ringworm of the scalp. They should be examined and treated.

4. Jiggers are a type of flea found in Africa. The female burrows under the skin to lay her eggs. As the eggs inside her develop, she swells up, making a bump the size of a pea on the skin's surface. The skin should be opened with a sterile needle and the pregnant flea removed. The tiny wound is dabbed with antiseptic and covered with a dressing. Wearing sandals is the best way to prevent jiggers getting into the feet.

5. Head lice used to be removed by shaving off all the hair and washing the head in paraffin. This was the quickest way but it was unkind, as the child was teased by his or her friends and made to feel 'dirty' or 'unclean'. There are many products which kill off lice. They also contain substances which melt the 'cement' so the nits are loosened and can be removed. Parents should examine their children's heads often as head lice is a common complaint in infant and junior schools. In the First World War, thousands of healthy young soldiers were killed, not by bullets, but by the diseases carried by body lice. A person infested with body or pubic lice should remove the hair by shaving, wash really well in the bath and then sprinkle DDT powder thickly over the whole surface of the skin. Clothes are covered in DDT powder, put into a bag which is firmly tied, and left for 2 hours. After this the seams are examined for any nits which may still be there. The treatment is repeated if necessary.

CONTROL OF BODY PARASITES

The most important control over body parasites is personal hygiene. Study the chart to find out if there is any way you can improve your own standard of hygiene.

The home also needs to be kept clean and fresh. Rooms should be well-aired, floors and windows washed, furniture dusted and polished. Bathrooms, kitchens and lavatories need daily cleaning. Rubbish should be wrapped up and kept in a closed bin till it is removed. Parasites breed in dirt and dust in dark and peaceful corners. Keeping your home clean and fresh is the best way of making sure they do not set up home with you.

In large cities, overcrowding is a serious problem. People have to share the same rooms, beds and furniture and it is quite difficult to keep up a high standard of personal and home hygiene. There is always a risk of infestation where people live in crowded conditions. Health visitors give advice on how to improve hygiene and get rid of body parasites.

Questions and things to do

Examine skin cells under the microscope. Draw and label what you see.

Draw the diagram of a section of skin and label it. Write about the function of each part labelled. Make sure you can read a thermometer quickly. Study a world map showing regions of high humidity.

1. Make a list of all the ways in which the body gains heat and loses heat.
2. What is meant by a 'constant' body temperature?
3. Explain carefully how the skin (a) keeps *in* body heat when we are cold and (b) lets *out* body heat when we are hot.
4. Explain how heat is lost from the body by sweating.
5. What is meant by humidity? How does it stop us cooling down?
6. Why is overcooling in babies so dangerous?
7. Make a list of other ways in which man helps to control his body temperature.
8. Why is hygiene of the skin so important for good health?
9. Do a project on skin hygiene.
10. 'The skin acts as a barrier.' Write about this sentence, explaining carefully what is meant by secondary infection.
11. Examine and draw a flea or a louse. Study the mouth parts.
12. Write a short account of any body parasite. Include (a) the damage it does and (b) the treatment and control.
13. Make a list of all the things which might cause body odour.
14. Why should wounds be cleaned with antiseptic and covered?
15. Different clothes are worn in different climates – and at different times of the year. Do a full project on the materials used for clothing in Britain in the summer and in the winter. Include facts about which materials keep in or let out body warmth, absorb sweat, are hard-wearing, easy to wash, quick-drying, inexpensive, and so on. Use pictures or illustrations to explain your text.

Chapter Ten

Getting information

Our brain 'looks after us'. It protects us from harm and makes sure everything is working well. To do all this, it has to know what is going on
  inside the body,
  on the surface of the body,
  outside the body (our surroundings are called our
  environment).
But the brain is tucked away inside the cranium, the skull, at the top of the body. It is a long way from the stomach, the legs, the skin. It is even further from our environment. There has to be a special system to get the information the brain needs, to find out what is going on inside, on the surface and outside the body. This special system is the **sensory system** and it is made up of **sense organs** and **internal receptors**.

THE SENSE ORGANS
These are on the surface of the body. They are for sight, smell, sound, touch, balance and taste. They have nerve endings which receive information. The nerve endings are called **receptors**. The information picked up by a receptor is called a **stimulus**. The stimulus is passed along a **nerve** to the brain. We see, hear, feel, and so on in the brain. The brain translates the stimulus into sight, sound, touch.

THE INTERNAL RECEPTORS
These are inside the body. They pick up information about what is happening inside us and pass it to the brain. The sort of information they get is that the bladder is full and needs emptying, that the stomach is empty and needs filling, or that a muscle is tired and needs to be rested. They also keep the brain informed about where the parts of our body are. This is very useful. We don't have to look to wash our ears, put food in our mouths, shrug our shoulders, or scratch an itchy place on our leg. Do these actions yourself and you will realize how this **sense of position** is useful. Write down the ways in which you think it is helpful.
The sensory system is extremely important for getting information and for protecting us from danger. Write

Fig 10.1

down some of the ways we could be in danger if we were blind, deaf, or had no feeling in our skin. Our sense organs are also precious to us as they give us great pleasure from our environment, such as listening to exciting music, looking at beautiful things, and so on. They are called the 'Gateways to the Brain'.

Sight

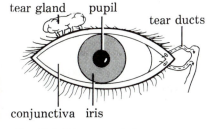

Fig 10.2 Diagram of the eye

PROTECTION OF THE EYES

The eyebrows stop sweat and rain from getting into the eye. We use them to show our feelings, pulling them upwards to show surprise and downwards to show anger or worry.

The eyelids close over the eyes about once every 6 seconds. Blinking keeps the eyes damp and sweeps any dust away from the front of the eye.

The eyelashes stop dust from getting into the eye. Tiny sebaceous glands secrete oil along the lashes. A **stye**, which is a small boil on the eyelid, is caused by an infection of the hair follicle.

The tear glands produce small amounts of tears all the time. Tears are a watery, salty, sterile fluid with enzymes to destroy germs. Tears are trickled down onto the eye, keeping it damp and clean. Find the tear ducts at the inner corner of your eye. The tears which have not evaporated drain down this canal into the back of your nose. When we cry, we produce excess tears which spill over the eyelid and pour down our cheeks. Work out (*a*) why our eyes get puffy and (*b*) why our nose runs, when we cry. We cry for many reasons, not only when we are sad.

The conjunctiva is a clear lining which covers the front of the eye and the inside of each eyelid. It protects the delicate front of the eye. If it gets infected, the eyelids swell and redden and the eye looks pink. This is known as 'pinkeye' and its medical name is **conjunctivitis**. We get conjunctivitis, like styes, by rubbing the eyes with dirty hands or drying them with an infected towel. It must be treated at once so that no damage is done to the eye and we do not pass the infection to other people.

MOVEMENT OF THE EYES

We only see the front part of the eye. It is important to remember the eye is nearly round and that most of the eyeball is deep in the skull, protected by the bony eye socket. Feel around your eye to find the rim of the bony socket. The eye is held in position by three pairs of muscles. Stare at the tip of your nose, then up to your eyebrows, out of the corners of your eyes, and all around you. You will feel the muscles pulling your eyes. Eye muscles work together, so *both* eyes are pulled in the same direction and we see an object with *both* eyes at the same time. What do you think happens if one of a pair of muscles loses its muscle tone?

A baby's eye muscles may not work together at first. It is quite usual for new-born babies to **squint**. If the baby is still *squinting* at 3 months, it must be treated or the squint will get worse.

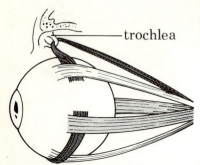

Fig 10.3 The three pairs of muscles which move the eyeball

Structure and function of the eyes

Our eyes give us a clear, detailed and exact picture of our environment. We can also judge distance, colour, brightness and darkness. The function of the eye is to let in light rays to the back of the eye, where they stimulate the nerve endings, the sight receptors.

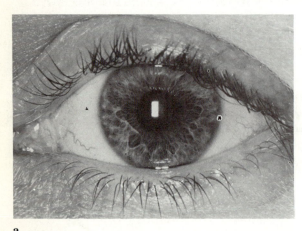

a

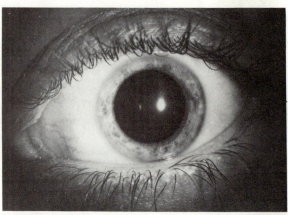

b

Fig 10.4 a The iris closes in the bright light
   b The iris is pulled back in dim light

The **pupil** is the part of the eye where light enters. The function of the **iris**, the coloured part around the pupil, is to control how much light gets into the eye. In dim light one set of muscles in the iris contracts pulling the iris back so that there is a very large pupil. In bright light the other set of muscles contracts, pulling the iris inwards so that there is only a very small pupil. In this way we get more light coming in when it is dark and we get protection from too much light when it is very bright sunshine or artificial light. We have no control over the muscles of the iris. They are involuntary. Sit in a dark room with a mirror and a torch. Switch on the torch and stare at one of your eyes. You will be able to see the iris closing, protecting the eye from the glare of the torchlight. Dark brown or black eyes have lots of pigment in the iris. Blue eyes and shades of green, grey and hazel have very little pigment.

There are three coats which make up the outside 'shell' of the eyeball:

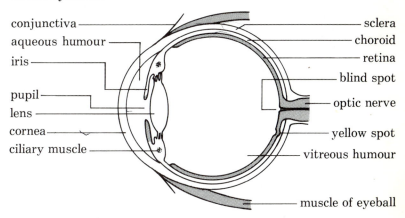

Fig 10.5 Keep checking with the text to make sure you understand where each part of the eye is

conjunctiva — sclera
aqueous humour — choroid
iris — retina
— blind spot
pupil — optic nerve
lens — yellow spot
cornea — vitreous humour
ciliary muscle —

— muscle of eyeball

THE SCLERA
The **sclera** is the tough outer layer which protects the eyeball. We can see it as the 'whites' of our eyes. It doesn't cover the pupil as we would not be able to see through it. Over the iris and the pupil it becomes clear and bulges out a little and is called the **cornea**. The cornea is very sensitive to pain. If anything comes near it, the eyelids blink involuntarily and the eyes are filled with tears. If the cornea is damaged or scarred we lose clear sight. It is rather like trying to look through 'crazed' or clouded glass. A damaged cornea can be replaced by surgery. The healthy cornea is taken from the donor after death. Thousands of people can get their sight back because other people donate their eyes to eye hospitals.

THE CHOROID
The **choroid** is the middle layer of the 'shell'. It has very dark pigment in it so that light coming into the eye isn't reflected away. This is why the pupil always looks black. (If you look inside a camera you will notice it is painted black.) The choroid has blood vessels to bring oxygen and food to the eye. As it comes to the front of the eye, one part is developed to form the iris. The other part of the choroid develops into the structure which holds the lens in place and allows movement.

THE RETINA
The **retina** is the very delicate inner layer, and is made up

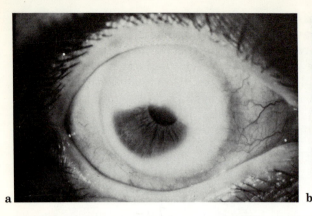
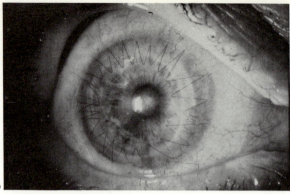

Fig 10.6 **a** A cloudy cornea
b The same after surgery. The stitches will come out in a few days

of millions of nerve endings that are sensitive to light. They are **light receptors**. There are two types of light receptors; **rods** and **cones**. Rods are used in the evening as they are sensitive to dim, low light. Cones are used in the daytime and in artificial light at night. They are sensitive to bright lights and colours. Why do you think we don't see colours clearly in the evening light? When we switch off the light at night, enter a cinema, or go anywhere from brightness to darkness, we can't see anything. For a short while we are **night blind**. This is because rods have a pigment called **visual purple** which fades in bright light. As we step into the dark, the rods re-form this pigment with the help of vitamin A. As it re-forms we can slowly make out shapes and objects and then we can see quite well. A few people don't have enough visual purple in the rods. They may be helped by eating more vitamin A foods (page 75). It is dangerous for such people to drive at night.

The retina is pressed against the choroid. But it may come away, after a blow on the head or in late middle age. This is called a **detached retina**. A very skilful eye surgeon can sometimes put it back into place.

The **yellow spot** is the place on the retina, directly behind the pupil and lens, where there are most cones and we see most clearly.

The **blind spot** is the place on the retina where the nerve endings of the rods and cones join up to form the **optic nerve**. The optic nerve passes from the back of each eye to the brain. As there are no nerve endings here, the eye does not receive light messages. This is why it is called the blind spot.

Colour blindness affects about 8 per cent of men but is rare in women. It is not serious as a colour blind person can usually *see* red and green, but cannot see *shades* of red and green. A rare type of colour blindness is not being able to see shades of yellow and blue. A completely colour blind person only sees grey, black and white. His sight is very weak. Colour blindness is inherited and thought to be the result of a fault in the cones.

Between the cornea and the lens is a watery liquid called the **aqueous humour**. The whole of the back of the eye is filled with **vitreous humour**. This is jelly-like. It helps to keep the eyeball firm and has dissolved oxygen and food in it.

Fig 10.7 a Front view of lens
b Side view of lens

a   b

THE LENS

The **lens** is directly behind the pupil. It is clear and thicker in the middle than at the sides. It is held in place by ligaments and has **ciliary** muscles for movement. When the eye is resting the lens is long and thin so that it can look at distant things. When the eye is looking at close objects the ciliary muscles contract, causing the lens to become shorter and thicker.

ACCOMMODATION

The function of the eye is to focus light rays on the retina. This means that as the light rays enter and travel through the eye they must be bent, **refracted,** so that a clear single image lands on the retina. Rays from a near object have to be bent sharply while those from a distant object need only slight bending. The light rays are bent by the curved cornea, the aqueous and vitreous humours, and the lens. The lens changes in shape to adjust the amount that the rays are bent so that the eye can focus on objects at different distances.

The eyes work much harder when we look at something near us.

The muscles of the eyeball pull the eye slightly inward.

The iris contracts so the light rays are narrowed.

The ciliary muscles shorten and thicken the lens to bend the light rays sharply.

The ability of the eyes to do these things, to adjust for near or far sight, is called **accommodation**. Can you think why the eyes feel tired after a lot of close study? They should be rested by looking at distant objects, such as trees and chimney tops and far-off views.

LONG SIGHT AND SHORT SIGHT

In normal sight, the light rays are bent to focus directly on the retina. Long sight is usually caused by the eyeball being small or by the lens not being able to become thick enough to bend light rays from close objects. We only see distant things clearly. After the age of 40 many people become long sighted and need glasses for near sight. Short sight is usually caused by the eyeball being too long. We only see near things clearly. Figure 10.8 shows how glasses can correct long or short sight. Copy these diagrams in your book and learn them.

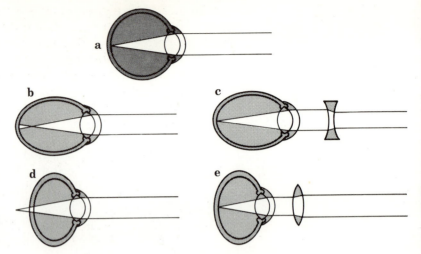

Fig 10.8 a Normal eye
b In short sight the light rays focus in *front* of the retina
c Short sight corrected with a lens
d In long sight, the light rays focus *behind* the retina
e Long sight corrected with a lens

CATARACT

As we get older the lens may become cloudy or split. Vision is weak and as the cataract develops over the whole lens, the person loses his or her sight. The lens can be removed and the person is given two sets of glasses, one for close vision and one for distant vision.

THE OPTIC NERVE

This is a cable of more than a million bundles of nerve fibres. Any severe damage to the nerve may cause loss of sight, as messages received by the retina cannot be passed to the brain.

The image which lands on the retina is upside down. It is thought the brain turns it up the right way for us.

Health and hygiene of the eyes

Anything which gets into the eye, such as dust or grit, is called a 'foreign body'. Extra tears are produced to try to wash it out. If it is a tiny speck, try to get it out with a clean tissue or handkerchief. Pull the lid away from the eye and look for the speck first. Then very gently lift it out with the corner of the tissue. It is very easy to damage the delicate conjunctiva and to infect it with germs. Never rub the eye, as you can push the speck into the membranes. Bits of steel, chips of glass and grains of sand must be removed by a doctor.

EYE CARE
1. All children should have regular check-ups on their eyes.
2. Always wear protective, shatter-proof goggles when doing dangerous work.

3. Any eye infection must be treated at once and care taken not to spread the infection to other people.
4. When doing close work, rest your eyes by looking into the distance every half an hour or so.
5. Make sure your work is well let; lighting should come from over the shoulder onto the work.
6. Avoid glare or dazzle in the eyes.
7. Don't use eye washes unless your doctor says so.
8. Foods with vitamin A help towards a healthy conjunctiva and against night blindness.
9. Any change in your vision or damage to the eye must be reported to your doctor.

BLINDNESS

In Britain, a blind person is someone 'unable to perform any work for which eyesight is essential'. It is important to understand that there are degrees of blindness, from having no sight at all to having what is called **partial sight**. There are many more partially sighted people than totally blind people. Each country helps its blind people in different ways. In Britain, children go to special schools for the blind and partially sighted.

Fig 10.9 Children at a school for the partially sighted. Notice the special equipment

Blindness in babies may be caused by the mother having German measles, which is called **rubella**, in early pregnancy. In poor countries and in tropical climates, there are many more blind people. Malnutrition in the pregnant mother and in the small child is a very common cause. Can you imagine the sadness of having a blind child because you cannot afford to buy enough food? Worm parasites, carried by flies in hot countries, are also the cause of millions of people losing their sight. Blindness also happens in very old age to some people.

Parents of blind babies need to know that lack of sight does not change the baby's intelligence. But blind children must have things explained to them so they can learn about what they can't see. If a blind child hears a dog barking, he cannot tell whether it is a big frightening dog or a small friendly one. Nor can he tell if the dog is held on a leash or is able to jump at him. This sort of problem can be got over by telling the child what is happening *before* it happens. Parents have to be the 'eyes' for their blind child, protecting him from fear by telling him what is going on around him. This helps the child not to be frightened of the world he cannot see, to try out new skills and to develop his mind and other senses to the full.

Smell

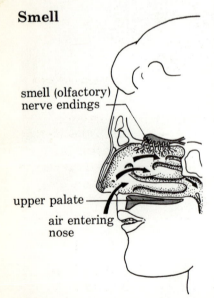

smell (olfactory) nerve endings

upper palate

air entering nose

Fig 10.10 The nerve endings for smelling

There are two small patches of tissue up in the back of the roof of the nose. These are the sense receptors for smell. They are made up of tiny cilia, bathed in a mucus layer. They join up to form the **olfactory nerve** which sends messages of smell to the brain. It is not clearly known how this happens but all smells are first dissolved in the mucus before the nerve endings can get the stimulus.

As we breathe in, air passes underneath the olfactory nerve endings. When we want a really good smell, we sniff, so the air is forced right up to the mucus layer. New earth, warm skin, old books and fresh rain are pleasing smells. Make a list of pleasant and unpleasant smells.

Smell helps us to recognize dangers, like escaping gas or burning food. It gives us more information about what is happening around us. But we only smell a new odour for a short while. This explains why we can be unaware that food is burning and yet someone who comes into the room can smell it immediately. Some people have a much better sense of smell than others. Smokers and people with heavy colds lose their sense of smell.

Hearing

If you talk to the person next to you, or bang a book on the desk, the air **vibrates** with **sound waves**. These have to travel into the ear and stimulate the hearing receptors so that messages can be sent to the brain and we can hear.

The flap of ear on the side of our head is called the **pinna**. It is the least important part of the ear and we can hear perfectly well without it. The main parts of the ear are buried deep in the skull and protected by bone. The ear can be studied in three parts – the outer ear, the middle ear and the inner ear.

THE OUTER EAR

The pinna is used for catching and directing sound waves into the ear. If your ears stick out, they will catch more sound waves than if they are flat against your head. Cup your hand behind your ear. What do you notice? The pinna has a good blood supply to keep it warm. It also may grow thick hairs, especially in men, to stop dust and small objects from getting into the ear.

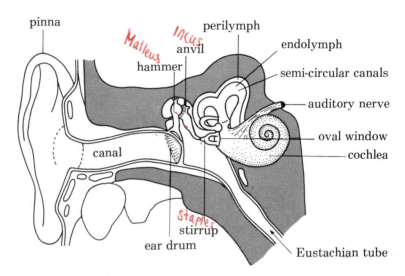

Fig 10.11 A diagram of the ear

The hearing canal, **auditory canal,** is a short tube leading from the pinna to the ear drum. It is lined with tiny hairs and secretes a waxy substance. Both these things help to trap any dust and dirt. At the end of the canal is the **ear drum**. It is a tightly stretched membrane completely closing off the outer ear. When sound waves travel down the hearing canal, they bump against the drum and make it vibrate.

THE MIDDLE EAR

There are three odd-shaped little bones forming a chain across the middle ear. They are called the **hammer**, the **anvil** and the **stirrup**, because it was thought they looked like these things. They touch one another. The hammer also touches the ear drum and the stirrup touches the **oval window**, another tightly stretched membrane. The vibration of the ear drum starts the three little bones vibrating, which cause the oval window to vibrate.

The middle ear is like an air pocket in the skull. It is sealed at each end by the ear drum and the oval window. The only way air can enter or leave is by the **Eustachian**

tube (page 50). The air pressure in the middle ear is usually the same as it is outside. If we go up or down in a lift, or an aeroplane, we may hear curious squeaks or go deaf for a short while. Swallowing or yawning 'pops' the ears. It makes the pressure in the middle ear the same as that outside again by forcing air up or down the Eustachian tube from the back of the nose.

THE INNER EAR

The inner ear is made up of strange looking objects, all bathed in a fluid, the **perilymph**. Find the **cochlea**, which looks like a snail and has the sound receptors in it. The sound receptors called the **auditory nerve endings**, are tiny hair cells bathed in another fluid, the **endolymph**. The vibrations of the oval window cause the perilymph to vibrate. This causes the endolymph to vibrate and the hair cells get the stimulus and send messages along the auditory nerve to the brain.

SUMMARY OF HEARING

Sound waves travel along the auditory canal.
They start the ear drum vibrating.
The vibrations are passed along the three little bones.
They start the oval window vibrating.
The fluid perilymph picks up the vibrations and passes them to the endolymph.
The auditory nerve endings are stimulated.
Messages are passed to the brain by the auditory nerve.
The brain translates the messages and we hear sound.
We can also hear through the bones of the skull. Vibrate a tuning fork and press it to your forehead. Vibrate it again and hold it between your teeth.

Health and hygiene of the ears

There is a saying 'Never put anything smaller than your elbow in your ear!' This means, of course, that you should never poke your ears or put anything in them at all. The ear flap should be washed daily to get rid of dead cells, dirt and any wax which comes from the canal. But the canal itself should not be touched. Some parents poke cotton buds into their baby's ears to clean them. This is not necessary and it is dangerous as the ear drum can be damaged. If the wax in your ears does become hard and you think it is making you a little deaf, it should be syringed out by the nurse at the clinic or by your doctor. Don't try to remove it yourself.

MIDDLE EAR INFECTIONS

These are quite common in childhood and are caused by infections travelling up the Eustachian tube from the nose or throat. Any infection of the middle ear *must* be treated by a doctor as there is a risk of long-term infection leading to deafness. The nose should always be blown gently, as any rough blowing may force mucus up the Eustachian tube into the middle ear. If you jump feet first into a swimming pool, hold your nose. This stops water being forced up it and getting into the Eustachian tube. Infections travel quickly between the ear, nose and throat so hospitals have a special E.N.T. department for treating them.

DEAFNESS

As with blindness, there are degrees of deafness; some people live in a completely silent world while others can hear with the help of a deaf aid. The most usual causes of deafness in the middle ear are (a) from an infection, (b) overgrowth of the little bones which stops them vibrating, and (c) excess noise. Damage to the cochlea and auditory nerve of the inner ear, which can be caused by noise, produces total deafness.

NOISE

Noise intensity is measured in **decibels**. A whisper is 30 decibels and normal conversation is 60. The maximum noise level for safe hearing has been put at 87 decibels. Short bursts of very loud noise or long periods of noise below the painful level can cause permanent loss of hearing. Fireworks, riveting, road drills, big guns and the engines of jet aeroplanes can cause deafness. Noise is also painful and irritating and more accidents happen where people are working in noisy conditions. Noise has been called a pollutant, as it is made by us and damages our health. The level of noise increases each year as we build and use more and more machines. How loud do you turn up your radio or record-player? Work out when you were last in silence, when you could not hear one single sound.

BABIES AND SMALL CHILDREN

A baby's hearing must be tested at intervals by health visitors, nurses at the clinic, or by your doctor. It is important to find out as quickly as possible if a baby has any hearing problems. This is because our speech depends on our hearing words spoken to us. During the first year of life we listened to the words spoken to us. During the second year we tried to copy the words. And so, very slowly, we learned to talk. A deaf baby misses the long time of listening to words while his brain is developing. He cannot learn what he doesn't hear, nor can he try to copy speech. We used to think some babies were born 'deaf and dumb'. They are not. They simply never learned to speak because they never heard speech. Babies with hearing difficulties should be fitted with deaf aids in their first year of life and spoken to as much as possible. They then have a chance of developing speech.

Deafness in new-born babies may be caused by the mother having rubella in early pregnancy. Small children may become deaf from diseases such as meningitis, mumps and middle ear infections. In industrialised countries, where some parents are not patient and kind to their children, deafness is caused by hitting the child on the head or shaking him violently. Deafness in old age is mainly caused by the auditory nerve not being able to pass the messages to the brain any more.

Balance

Did you spin around like this when you were a child? What happened when you stopped? Did you feel giddy or fall over? This could have happened because you upset your sense of balance.

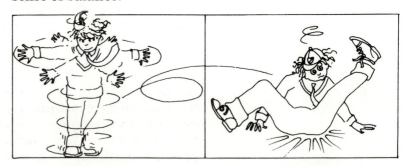

Fig 10.12

three semicircular canals at right angles

Fig 10.13 The semi-circular canals

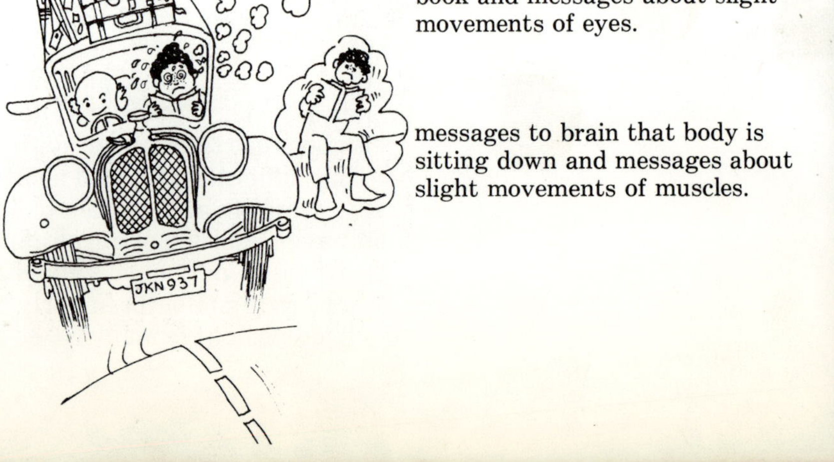

Our sense of balance has been called our 'sixth' sense and the organs which control it are in the inner ear. Look at the objects of the inner ear we haven't yet studied. There are three **semi-circular canals** attached to a large base with a smaller one underneath. All these organs control our sense of balance by passing messages to the brain about our position; whether we are spinning round or tilting our head, or lying down, and so on. The organs are lined with nerve endings covered with fluid, the endolymph. When we move our head, the endolymph moves and presses on the nerve endings. So messages are passed

messages to brain from endolymph swinging from left to right as car turns corners.

messages to brain from words in the book and messages about slight movements of eyes.

messages to brain that body is sitting down and messages about slight movements of muscles.

Fig 10.14

to the brain. The brain works out the position of the body and sends messages to the muscles.

Our sense of balance is also helped by our sight and by the internal receptors in the muscles. It is much more difficult to balance on one foot with your eyes closed than with them open. It takes time for small children to develop the proper working, **co-ordination**, between sight, internal receptors and organs of balance. They fall over quite often.

Study Figure 10.14. Such a conflicting set of messages confuses the brain. We feel dizzy and sick. If we were standing, we might fall over, as the brain cannot sort out which messages to send back to the body to keep it upright.

It now becomes easy to understand why we get travel sick and why it is not wise to read in a moving vehicle.

Taste

Our sense of taste seems the least important of the senses. It doesn't protect us from danger, as food which is bad does not always taste bad. And it doesn't give us much information about our surroundings. But it does give us pleasure as we enjoy eating our favourite foods.

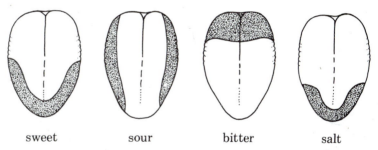

Fig 10.15 The main areas of taste

| sweet | sour | bitter | salt |

It is thought there are about 9000 taste buds on the tongue, and there are also some on the linings of the cheek and the pharynx. We have four main tastes. They are sweet, bitter, sour and salt. All other tastes are mixtures of these four. Study Figure 10.15 to see where the different tastes are on the tongue. Do your own tests with strong-tasting food to find out where your different taste buds are. Taste buds are tiny hair cells in small pits on the tongue. These hair cells are the nerve endings, the **taste receptors**, which pick up the stimulus and pass it to the brain. It is not clearly understood how this happens. But it is known that the 'taste' in food must first be dissolved in saliva before the buds are stimulated.

Our sense of smell adds to our taste. We sniff a delicious cooking smell and we can almost taste the food before we put it in our mouth. When we have a heavy cold, we may think we have lost our sense of taste. We haven't. We have only lost our sense of smell. We are left with our sense of taste which is not very strong on its own. Try this out for yourself by holding your nose before and during eating.

133

Touch

The skin is the sense organ of touch (page 103). It is thought there are five different types of receptor in the skin:
1. for light touch;
2. for heavy touch, which is pressure;
3. for pain;
4. for heat;
5. for cold.

We have studied the importance of the receptors for heat and cold. The other receptors are equally important. Pain, which we think of as an awful thing, is very useful for teaching us to avoid danger and damage. If we had no pain receptors we wouldn't know when things were going wrong. Without toothache, we don't know a tooth is decaying. Without pain, we might get into a bath of boiling water and damage our tissues terribly. Pain is nearly always a warning that something is going wrong or that we are in danger of something going wrong.

Slight pain, such as headache or indigestion, will usually go away once we have rested or slept or stopped eating the wrong foods. There is little need to take pain-relievers such as aspirin or stomach powders, as the body can put things right by itself. Severe pain should be reported to a doctor if there is no clear reason for it, such as a cut or an attack or cramp, and if it doesn't go away quickly.

Our touch receptors can give us great pleasure. We may like the feel of silky clothes or snuggly woollen blankets against our skin. We like being stroked and touched by the people who love us, as we like to caress them in return. We enjoy feeling smooth wood, cool stone, polished surfaces. We may like feeling sun on our bodies, soft wind in our hair, or rain on our faces. What special things do you like? Getting pleasure is an important part of our lives. It helps us to be happy and cheerful and hard-working. It gives us something to look forward to and to remember, to look back on with delight. It is a large part of our sex drive which we will study later.

Questions and things to do

1. Examine carefully a complete eye from a bullock, pig or sheep (from a slaughter-house). Find the eye muscles and the optic nerve. Draw the complete eye.
2. Open the eye, starting at the optic nerve. Look for and identify all the structures in your diagram. Draw the opened eye and label it clearly.
3. Remove the lens, wash it, feel it. Then place it over some print.
4. *To find your blind spot* (Figure 10.16). Hold the book about 60 centimetres away. Close the left eye and stare at the cross. Slowly bring the book near your face.

The dot will seem to disappear when its image falls on the blind spot.

Fig 10.16

5. *To find which eye is used more.* Hold a pencil at arm's length, in line with a distant object. Close one eye and then see if the pencil 'moves'. Close the other eye and do the same. If the pencil 'moves' more when the right eye is closed, it means your right eye is used more than your left. And of course, the other way round.

6. *To test your hearing.* Blindfold your eyes and put a plug of cotton-wool in one of your ears. Ask your partner to ring a bell or blow a whistle from different parts of the room. Each time you hear the noise, point in the direction you think it comes from. Then repeat the test, plugging up the other ear. Your partner will tell you how accurate your hearing from each ear is.

7. Try to work out a simple test to show that a patch of skin on your shoulder might have less sense receptors for gentle touch than the skin on one of your fingers.

8. Draw the structure of the ear.

9. Do a project on deafness or blindness. Visit a school or training centre to get direct information. Write to any one of the many societies there are in the world to help blind and deaf people. You may be interested to join the International Noise Abatement Society.

10. Draw a large labelled diagram of the eye. Explain as simply as you can the function of each part you have labelled.

11. Draw a large labelled diagram of the ear. Explain as simply as you can how the different parts help in hearing.

12. Explain why an infection in the throat may cause you to become slightly deaf.

13. What things help us to have a sense of balance?

Chapter Eleven

Dealing with information

The nervous system is made up of the **brain** and the **spinal cord**, the **nerves** with their **nerve endings** and the **autonomic nervous system**. It is a huge communication system which spreads right through the whole body. It deals in messages. Messages are brought *to* the brain. They are sorted out and dealt with *in* the brain. New messages are passed *across* the brain and sent *out* to the body. Very simple messages don't go to the brain. They can be dealt with by the spinal cord.

The nerve cells which carry the messages through the nervous system are called **neurones**. Neurones are different shapes and sizes depending on where they are and what they have to do. But they all have the same basic structure. Study the single neurone in Figure 11.2 and find:

Neurones

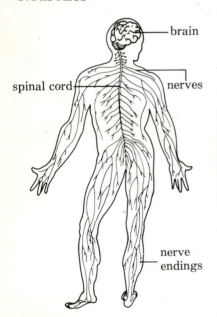

Fig 11.1 The nervous system

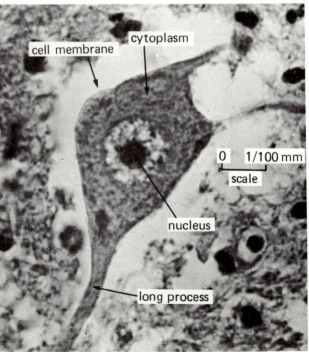

Fig 11.2 A neurone

the large nucleus,
the **cell body** around the nucleus,
the **nerve fibres**, which are the little branches coming from the cell body.

The nucleus and cell body decide what is to be done with the message. The nerve fibres bring a message in, or pass a message to another neurone, or take a message away from the cell body. So you can think of neurones as being one of three main types: **sensory neurones** for bringing the message in, **connector neurones** for sorting the message out in the brain or spinal cord, and **motor neurones** for sending messages out to the body.

Sensory	*Connector*	*Motor*

Fig 11.3 Jim's foot is aching. The message is picked up by the sense receptors in Jim's foot and passed along the nerve to his brain

Fig 11.4 Jim decides his shoe is too tight. The message is worked out in Jim's brain. It is passed across his brain by many connector neurones till it gets to the motor area

Fig 11.5 Jim takes off his shoe and rubs the sore place. The motor cells of the brain send messages to his eyes, muscles and internal receptors to take off his shoe

FACTS TO REMEMBER
1. The message we get is called a **stimulus**.
2. A stimulus is picked up by a nerve ending called a **receptor**.
3. The correct name for a message which passes into, across, or from the brain or spinal cord is a **series of impulses**.
4. Impulses travel very quickly by chemical and electrical changes in the nerve fibre.
5. A sensory nerve fibre can be long, e.g. from the tip of your finger to the spinal cord, or it can be short, e.g. the optic nerve is close to the brain.
6. Connector neurones are in the brain or spinal cord. They have short fibres and their job is to form a link

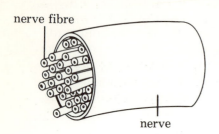

nerve fibre

nerve

Fig 11.6 A diagram of a nerve

Nerves

A reflex action

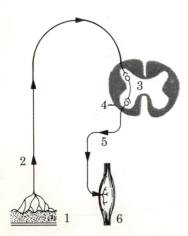

Fig 11.7 Check this diagram of a reflex arc with the points in the text

between the sensory and motor nerve fibres. The ends of the connector neurone fibres never quite touch the ends of the other nerve fibres. The small gap is called a **synapse**. Impulses or messages 'flash' across the synapse.

7. A motor nerve fibre has nerve endings called **effectors**. They cause an effect. Effectors make the muscles contract or relax. They make the glands produce more or less secretions.

8. A motor nerve fibre can be long, e.g. from your spinal cord to the muscle in your foot, or it can be short, e.g. from your brain to the hinge muscle of your jaw.

A nerve is a collection of nerve fibres, all bundled together. Nerve fibres are separately wrapped in a white **sheath** for protection. Around the outside of the nerve is another sheath for protection. Nerves look like white cables. Large nerves carry both sensory and motor nerve fibres. They split up into smaller and smaller nerves. Very small nerves carry either sensory nerve fibres or motor nerve fibres but not both.

This is an action we make which is very fast and very simple. It is an involuntary action as it is not controlled by the mind. It is 'inborn' or 'instinctive' and is usually for our immediate protection. Sudden blinking, sneezing and coughing are nearly always reflex actions. We can watch the opening and closing of the iris, which is another reflex action. Try the knee jerk reflex action now. Cross one knee over the other, and let the crossed leg hang limply. Tap the tendon just below the kneecap with the side of your hand. Your leg will shoot out in a sudden kick. You have no control over this kick. (You may have to try this a few times before you find the exact place to tap.)

A reflex action is important because it happens so quickly. The message is picked up and flashed in to the nearest part of the nervous system. *It does not have to go to the brain.* It jumps across the connector neurones and motor messages are flashed out to the muscles. Do the knee jerk once more to realize how fast the action is.

Imagine you have touched a hot plate and immediately snatched your hand away.

1. Receptors in the finger are stimulated by heat and pain.
2. Messages are flashed up the sensory nerve fibres of your arm to the spinal cord.
3. The messages are passed across the spinal cord by the connector neurones.
4. New messages are started in the motor neurones.
5. Messages are flashed down the motor nerve fibres to muscles in the arm and shoulder.
6. The effector nerve endings in the muscles receive the

message to contract and your hand is quickly pulled away.

From the diagram, you will see the whole action forms an arc. This is called the **reflex arc**, and the path it follows is called the **nervous pathway**.

A conditioned reflex

Imagine you hear the bell for lunch. Without thinking, you pack away your books and go to the lunch room. You are talking with your friend and not aware of what you are doing. When you sit down in the lunch room you may think 'Yes, I am here. But I don't remember how I got here. I don't remember the details of each action I took after I heard the bell.' A lot of our behaviour is learned behaviour. We have packed up books and gone to lunch so often that we don't need to think about what we are doing. Nervous pathways have been laid down in our nervous system and we follow them. This is very useful because it leaves our brain free to take in and deal with new experiences. Think how slow our lives would be if we had to concentrate on every little thing we do.

A lot of our behaviour is learned or **conditioned** behaviour. A **conditioned reflex** is an action we have learned and repeated so often that each time we receive the stimulus we react in the same way without thinking about what we are doing. Pavlov, a famous Russian scientist, did this experiment with dogs.

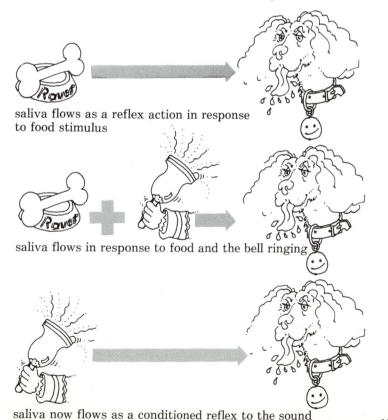

saliva flows as a reflex action in response to food stimulus

saliva flows in response to food and the bell ringing

Fig 11.8

saliva now flows as a conditioned reflex to the sound of the bell alone

Why do you think showing the food and ringing the bell had to be repeated many times? Work out one simple conditioned reflex action which you do.

Most of our conditioned learning happens when we are very small. Watch a baby taking its first steps. You can almost 'feel' the messages being sent from the brain to the muscles. Each new skill we learn has to be practised over and over again before it forms a nervous pathway and becomes a part of our behaviour. Make a list of some of the skills a baby learns in its first three years. **Habits** are conditioned reflex actions and are also learned when we are young. We learn habits of thinking, feeling and behaving from our parents and the people in our home. Some of these habits will stay with us for the rest of our life. You can understand how very important parents are in helping a child develop habits for a healthy, active and happy life.

Remember:

A reflex action is inborn and is not controlled by our conscious mind.

A conditioned reflex is learned by our conscious mind. After the learning period we do not need to use our conscious mind when we repeat the action.

We can control a conditioned reflex action by using our conscious mind.

The brain

The brain is a soft, greyish lump and looks like a large, wrinkled walnut. It weighs about $1\frac{1}{2}$ kilograms and is made up of about 15 000 million brain cells. It's difficult to think of 15 000 million of anything, though we have all these brain cells to think with! Even though brain cells are packed closely together, they do not touch. Messages must 'flash' across synapses from one neurone to the next. Parts of the brain are bathed in fluid, **cerebro-spinal fluid**, which brings oxygen and nutrients to the cells. Three parts of the brain we will study are the **cerebrum**, the **cerebellum** and the **medulla**. Find them on Figure 11.9.

THE CEREBRUM

You can see this is very large and covers most other parts of the brain. It is also very folded which allows room for far more brain cells. There is a deep groove in the middle, which almost divides the cerebrum into two halves. The left side of the cerebrum controls the right side of our body, and the right side controls the left side of the body.

The cerebrum controls our conscious acts.
It receives messages from the sensory organs.
It translates these messages.
It sends out new messages to the muscles and glands.
It deals in things we don't completely understand such as memory, intelligence, thinking, judging, decision making, conscience, emotions and imagination.

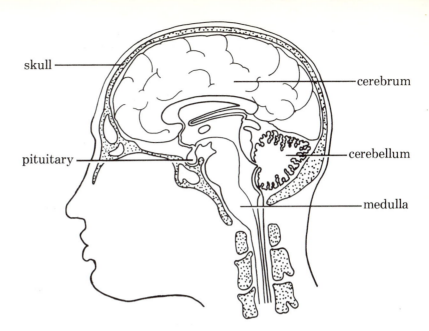

skull

cerebrum

pituitary

cerebellum

medulla

Fig 11.9 A diagram of the brain

THE CEREBELLUM

This means the 'little brain'. If you look at Figure 11.9 you will realize why it got its name.

The cerebellum deals with our balance and sense of position.

It deals with muscle tone and the smooth contracting and relaxing of pairs of muscles.

It deals with the co-ordination of our body and our movements. This means it puts together all the messages it receives and sends new ones to the muscles so our actions are correct and graceful.

THE MEDULLA

This is cone-shaped and forms a swelling at the top of the spinal cord.

It controls the vital activities such as respiration, digestion, heart rate and temperature control. It works at an 'unconscious' level as it must continue to work while the conscious part of the brain is asleep.

It controls reflex actions which are also vital activities. For example, if the swallowing reflex stopped (page 51), saliva would go down into the lungs and we would drown.

There are many other parts of the brain with important functions but it is not possible to study them here. However, what is most important to understand is that the brain acts as a huge co-ordinating centre. Though we know there are special centres for special functions, the brain acts as a whole, protecting and controlling itself and the body.

FACTS TO LEARN

1. The brain gets messages, as a series of impulses, from all the sensory organs of the body.

2. It translates these messages and sorts them out.
3. It passes the messages across connector neurones to the parts of the brain which deal with them.
4. Some messages are stored as information to be used in the future.
5. It sends out new messages to the muscles and glands along the motor nerves, or more of the same messages if they were the correct ones.
6. Parts of the brain work at our conscious level and parts work at our unconscious level.

The spinal cord

The spinal cord runs from the bottom of the brain to the second lumbar vertebra (page 10). It is a very simple part of the brain. In the centre is a small canal for the cerebro-spinal fluid which brings oxygen and nutrients to the neurones. The **white matter** you can see on the outside is made up of nerve fibres and nerves covered in their white protective sheaths. The **grey matter** which looks a bit like a letter H is made up of the cell bodies and nuclei of the neurones. The function of the white matter is to pass messages between the brain and the body. The function of the grey matter is to deal with reflex actions.

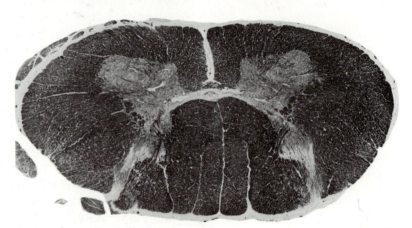

Fig 11.10 A section through the spinal cord

Protection of the brain and spinal cord

Which bones protect the brain and the spinal cord (page 9)? However, the skull does not touch the brain. Between the skull and the brain are the **meninges** and the **cerebro-spinal fluid**.

THE MENINGES
These are linings of the brain. There are three of them and they are made of membrane tissue. The outer lining is a tough membrane which covers the inside of the hard bony skull. The middle lining is delicate and web-like and fluid can pass through it easily. The inner lining wraps closely around the soft cells of the brain itself.

The linings have a rich supply of blood vessels. Brain cells need a great deal of oxygen as they have such a vast amount of work to do. In fact, brain cells die within 3 to 5 minutes if they do not get oxygen. Once a brain cell dies, it cannot be replaced. A new one doesn't grow in its place. It is vital that there is a rich supply of blood vessels bringing fresh oxygen to the brain all the time.

THE CEREBRO-SPINAL FLUID

This is a clear fluid, rather like plasma, between the middle and inner meninges. It acts like a water-cushion for the delicate brain cells, protecting them from any bumps or knocks and from the hard bony skull. It also brings oxygen and food to the brain and removes the waste products.

The spinal cord is also wrapped in meninges and is supplied with cerebro-spinal fluid. Work out the reason for the name of the fluid.

The autonomic system

This is a very important part of the nervous system. It lies outside the brain and spinal cord, though it links up with them. It controls the rate of our heart beat, respiration and so on and sends messages to all the vital organs and glands to speed up or slow down, when it is necessary. It works in two parts, one part controlling the action of the other, so that between them the internal workings of our body are kept at a proper **steady state**.

It is sometimes called the 'flight or fright' system as it also helps us when we are in danger. Remember the last time you were really frightened? How did you feel? Or imagine you are alone. It is dark. You hear sounds, scufflings, creakings or heavy footsteps. You are terrified. As your fear rises, you can almost 'feel' your body getting ready to run away and hide or to tense up ready to fight.

Fig 11.11 This man is 'keyed-up' to run a race. See page 163 for more information

143

The iris is pulled back so you can see more clearly.
The salivary glands stop secreting saliva.
Your breathing rate gets faster and faster.
You can hear your heart thumping as the heart beat
speeds up.
Digestion in the stomach and small intestine slows down.
The liver sends out extra simple sugars to the muscles.
Peristalsis stops.
Control of the bladder and rectum is relaxed.
Blood is drawn away from the surface of the body and
other places and sent to the lungs, heart and muscles.
These are some of the things which happen when you think
you are in danger. Your body is now in a state of 'Red
Alert', ready and prepared to run or fight, whichever is best
for you. Re-read this list and tick off the things which will
help you to act quickly. Study the things which are left on
the list. You will realize that none of them would help you
to get out of danger. So the autonomic system has very
cleverly slowed down or stopped their action. The extra
energy is used by the liver, lungs, heart and muscles.
Explain clearly in what ways each of these four organs will
help you out of danger. Why do you think some small
children, and a few adults too, pass water or wet
themselves by accident when they are excited or
frightened? Write down why you get that sinking feeling in
your stomach when you are in trouble.

Now, imagine the danger is over. You can't stay all
keyed up for action when there is no need for it. The other
part of the autonomic system begins working, slowing
down some things and speeding up others till the body
returns to its normal steady state. Write the list of things
which happen to return your body to its normal internal
steady state.

Remember:
The autonomic system is self-governing; we have no
control over it.
It is linked to the brain and spinal cord.
It lies just outside the spinal cord and sends messages to
the organs and glands.
It works in two parts, by speeding up some actions and
slowing down others.
Its function is to keep the body working at a proper
steady state.
In danger it gets the body ready to fight or run away.
After danger it returns the body to its normal steady
state.

Disease and damage to the nervous system

BRAIN DAMAGE
Sadly, about 4 in every 1 000 babies are born with brain
damage. There are many different reasons for this and not
all of them are known. Think of how complicated the

Fig 11.12 A child with cerebral palsy

nervous system is, how delicate neurones are, and how fragile a baby is before, during and just after birth.

There are different degrees of brain damage and some babies are only mildly affected. The correct name for brain damage is **cerebral palsy**. If the nerves going to the arms or legs are damaged, the muscles are not able to make voluntary movements smoothly and may become twisted and drawn. If they can make no movements at all, they are **paralysed**. Damage to the parts of the brain dealing with intelligence can cause a child to be backward. People with brain damage are not mentally ill. They are **mentally handicapped**.

Children with cerebral palsy can be helped by special treatment to train their muscles and to get some control in their brains. Severely backward and paralysed children are usually kept in special homes as they cannot do anything for themselves. The people who look after them need to be loving and have endless patience and understanding. Children with slight brain damage can live at home and go to schools for the educationally sub-normal. These schools also help children who need extra time and care to develop their learning.

CONCUSSION

This is a different type of brain damage from a blow on the head or a fall. The person loses consciousness and is not aware of himself or the things around him. When he becomes conscious again he feels dizzy and sick, has a headache, and must lie down to rest and sleep. (The hero in a film who can get up after he has been 'knocked out cold' and go on fighting does not exist in real life!) Concussion is always serious as there is a risk of brain damage, blood clots forming and many other troubles. Boxers can die from brain damage caused during fights. Any person who loses consciousness after a blow on the head should be taken to hospital.

EPILEPSY

This is a disorder of the brain's function. For a short while the epileptic person may lose conscious control. He or she will fall down, parts of the body may jerk or stiffen, the tongue may be bitten. This is the most serious form of epilepsy. There are other, much milder forms. We used to be frightened of epilepsy because we thought the person having an attack was 'mad'. This is not true. Many people who suffer these brief fits are more sensible and intelligent than other people. There are now marvellous drugs which can control epilepsy.

MENINGITIS

As you can guess from the name, this is an infection of the meninges. It is a very dangerous disease, especially in children and teenagers.

ENCEPHALITIS

This is an infection of the brain itself. Many different viruses may cause this disease and it can be infectious. The symptoms are that the person has headaches, sickness, drowsiness, blindness and may become unconscious.

POLIO, RABIES AND MULTIPLE SCLEROSIS

These are other diseases of the nervous system you may find it useful to study.

PARALYSIS

This is loss of movement in any part of the body. It is most usually caused by brain damage or brain infection. But it can be caused by an accident; if a large nerve is cut or **severed** right through, no messages can be passed either way. When a large nerve in the upper arm is severed, the lower arm is paralysed as the muscles do not get messages to make it move. (Smaller nerves are sometimes able to re-grow.)

Anaesthetics

These are drugs which are used to 'deaden' the nerves so we don't feel pain during an operation.
1. A **general anaesthetic** is breathed in through a mask and puts us into a deep sleep. It deadens the cerebrum and the motor and sensory nerves and is used in large operations.
2. A **spinal anaesthetic** is injected into the cerebro-spinal fluid in the spinal cord. It deadens the sensory and motor nerves below the place of the injection. The patient is wide awake but feels no pain.
3. A **local anaesthetic** is injected into a sensory nerve

and deadens the receptor nerve endings. It is very often used by dentists.

Sleep

Our minds are hungry for new facts, new information, new experiences. Each day we take in lots of new messages, some of which are quite difficult for us to understand. In the same way that our bodies need time to rest and recover, so our minds need time off to rest and sort out all the new information.

It is known that we can go for longer without food than we can without sleep. It is also known that sleep is vital for the health of our minds. But exactly why sleep is so important is not fully understood. Why do we spend one-third of our lives asleep? What do our dreams mean and why are some of them so frightening? Why do we all dream but usually forget what our dreams are about?

There are at least two kinds of sleep. In one, the growth hormone is released, new cells are built and worn-out cells are removed. In the other kind of sleep we dream. Our eyes make rapid eye movements behind the closed lids. It seems likely that our brain is sorting out the information it got during the day. We need both these kinds of sleep, which happen in turn right through the night.

During sleep most of our body activities slow down. The body temperature drops, the heart rate and breathing are slower, our muscles relax and the whole body rests. We wake up feeling full of energy and ready to meet the new day. Each person needs a different amount of sleep. The average for an adult has been worked out at 7 hours and 20 minutes. Teenagers need more than that and babies

Fig 11.13 Hogarth's famous picture of an asylum

sleep an average of 16–20 hours a day. We think babies sleep longer than this because they lie quietly for some of the time and we imagine they are asleep. Some people find it easy to sleep anywhere; others need a comfortable bed in a quiet, dark room.

Mental health and development

Until quite recently, people who suffered from any form of **mental ill-health** were thought to be 'touched in the head', 'crazy' and 'insane'. Their families were ashamed of them and shut them away in mental homes. People used to visit these **asylums** as a Sunday outing and pay money to watch the **lunatics**. They would laugh and jeer at them, treating them as if they were more like animals than human beings. 'How cruel', we think now. But we are still more ready to understand sickness of the body than of the mind. We are still kinder to an illness we can see than one we can't see. Can you think of any reasons for this?

Nowadays we know there are certain disorders of the mind which may cause the person who is mentally ill to be dangerous; either to himself or to other people. Because we don't know enough about these disorders, we keep dangerous people in mental hospitals or prisons for the criminally insane, for their own and for other people's safety. They are called **psychopaths**. Sometimes they can be cured of their disorders.

STRESS AND MENTAL ILL-HEALTH

We are going to study the problem of people who suffer from 'nervous breakdowns', 'nervous exhaustion', or 'anxiety states'. There are many different names for these sorts of mental illnesses. We will use the term mental ill-health. People who suffer from this become so deeply unhappy, or so anxious or disturbed, that they are not able, *for a while*, to lead normal healthy lives. Mental ill-health is increasing, particularly in industrialised countries, and one of the causes of this is the way in which we cope with stress.

Stress

Life is quite difficult for all of us at times. We feel unloved, we feel insecure, we are afraid of failing. We don't understand why other people seem to have a better life than we do. We get angry, jealous, spiteful and hateful, as well as pleased, happy and wild with excitement. We have to cope with all these feelings and this can be difficult. It is hard to learn not to hit out when our feelings have been upset; to control ourselves so we don't damage other people. We have to become social; doing some things we don't want to do and not doing other things we do want to do. All of us have to work out some sort of balance between our own feelings and those of the people in our world. Sometimes this causes us **stress**. Stress is different things to different people. Some people suffer stress from outside themselves – the world is too noisy, too full of machines, too polluted, too fast, too slow, too overcrowded, too lonely. Write down a list of things you think might cause you stress.

Some people are not able to find ways to cope with stress and become mentally ill. Other people enjoy stress, and because of it they work harder and longer and lead happy healthy lives. Most of us just cope with it as best we can and don't let it bother us too much. Why do some people break down under stress while other people don't? There are many different answers to this question but we are going to study only one of these answers. To do this we must understand how a person develops mentally and *emotionally* (in his feelings) so we must go back to the beginning of life.

Fig 11.14 Edvard Munch's *The scream*

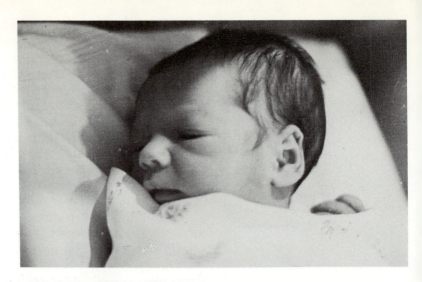

Fig 11.15

THE BEGINNING

We can't think of a baby's mind as being a blank. From the moment he starts life in his mother's womb, he is a mixture of things he has got or **inherited** from his parents. These things will control quite a large part of the sort of person he is likely to develop into. But equally important are the things which happen to him after his birth as a result of his **environment**, A baby's environment is made up of his parents, the home he lives in and the family and friends around him. Let's take a simple example of the way in which what we inherit and what we get from our environment help to make up the sort of person we become.

Baby Smith inherits a chance to be clever and pass exams easily.

Baby Smith's parents don't believe in education. Baby Smith isn't given crayons or paper and there are no story books in his home. When he goes to school he doesn't learn anything because his parents have told him learning is a waste of time. As he gets older, he plays truant and spends his time getting into trouble because he is bored. He leaves school as soon as he can and has to take a job which doesn't interest him at all. He goes from job to job till he finally stops working. It doesn't seem likely he will have a very happy future as he is fed up with himself and the world he lives in.

This is an over-simple story but it will help you to understand how our environment can affect part or the whole of our future lives, and how it is very important in shaping the sort of person we will be.

THE ENVIRONMENT

A baby has to be helped to develop his mind and his feelings. The people who do this are his parents or whoever is looking after him. Nowadays most parents are taught how to feed, clothe and keep a baby warm, dry and clean so his *body* develops well. But they are not taught how to look after his mind and his feelings. This is because we can't see a baby's mind and we can't see his feelings so we forget about them. Or we think they are not important. 'He's only a baby,' we say, 'he can't really feel or think.'

Learning 'bad' experiences

A baby is helpless. He has to lie and wait for the things he needs to be given to him. He is **dependent** on his parents for everything he needs. When he is hungry, cold, lonely or bored he feels 'bad'. He cries with anger till he is given what he needs. If his parents make him wait too long, his crying changes to fear and pain. He learns to be frightened of his mother and father and of the world he is in. He has no idea of time. He can only understand he is in some terrifying place where no one will come and help him. He is learning his first bad experiences; he is suffering from stress.

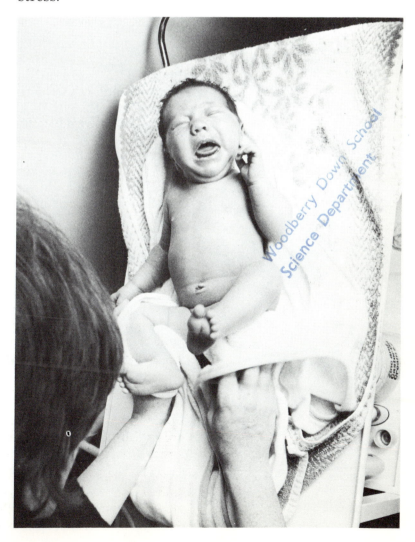

Fig 11.16

Learning 'good' experiences

A baby is dependent for a long while. If his parents go to him when he cries and give him what he needs, he feels 'good'. He begins to understand he is in some safe place where people come and help him and give him what he needs. He is learning his first good experiences; that he can cry, he can feel stress, but that it is quickly taken away.

While a baby is dependent, it is vital for the healthy development of his mind and feelings that he feels 'good'. A baby has as much stress in his life as we have in ours. We don't think of his stress as painful because it doesn't hurt us to wait, and we don't feel frightened by the same things. But it *is* painful for the baby, and part of the parents' most important work is to protect their baby from stress whenever they can.

Conditioned learning

We know that once we have learned to walk we don't forget how to do it. We don't even have to think about it as our nervous pathways are set for walking. In something like the same way, a baby or small child develops nervous pathways about his mind and his feelings. 'Bad' feelings may make him think of himself as bad; he is unloved, not wanted, useless and a nuisance. He may think of his parents as 'bad'; not to be trusted and able to give him pain and loneliness. He may think of the world as 'bad'; a cruel place, full of wars and wickedness. As he grows up he may get over these feelings, but when he is under stress his old fears will return and there is a chance he will suffer mental ill-health.

Babies and small children who experience 'good' feelings think of themselves as loved, wanted, safe and secure. They

Fig 11.17 A happy and secure child

trust and believe in their parents. They learn more quickly and are eager to try out new experiences. They find the world an exciting and happy place to live in. They can cope with stress.

This is a very brief outline of the way in which parents can either help a child to develop into a mentally healthy person or help him to develop into a mentally **insecure** person. It has only recently been discovered that the first years of life are very important for a person's mental growth and healthy development. Bringing up a baby isn't as simple as we would like it to be. It might have been simple long ago but it isn't simple any more. This is because we have become very complicated people living in very complicated societies. All parents need to learn as much as they can about the development of a baby's mind and feelings *before* they start their families.

REMEMBER

There are many different causes of mental ill-health. Stress is only one of these causes.
The first years of life are the most important for mental health.
Parents must protect their babies and small children from unnecessary stress.
Mental ill-health can be treated. Patients get better quite quickly, as they do from an attack of physical ill-health.

Other dangers to the nervous system

The nervous system is very complicated and any damage to it can have a serious effect on our health and on our enjoyment of life. In spite of this, many people risk damaging it by drinking too much and by the misuse of drugs. So it is most important to understand the dangers involved.

ALCOHOL
It is important to remember that alcohol is both a poison and a dependency drug.
It is a poison
It upsets the working of the brain cells because it slows down and muddles the messages the brain has to deal with. People don't often die of alcohol poisoning because they lose consciousness first. The brain goes into a deep sleep, the heart beat and breathing rate slow right down and the body temperature drops sharply. A person in this state of drunkenness is dangerously ill.
It is a dependency drug
This means that the body and mind build up a *need* for alcohol and become *dependent* on it. This doesn't happen suddenly. It takes a long time, so the person is not aware of the danger. He or she goes on drinking, believing it is easy to stop, until it is too late. When a person is dependent on alcohol he or she is called an **alcoholic**.

Fig 11.18

WHY DO PEOPLE DRINK?

Drinking alcohol is a part of our social life. We may be asked out to tea or we may be asked out for a drink. We go to restaurants and we go to public houses. It is quite usual for people to meet their friends for a drink and a chat. A glass or two of beer helps them to relax, to unwind after a hard day's work, to forget their worries and problems and enjoy themselves. It would seem that a little alcohol doesn't do us much damage, and if it helps to cheer us up it does quite a bit of good.

But – and when you are thinking about drugs there is always an extremely important 'but' – any drug which changes the way we behave has its own in-built danger. For example, alcohol, one of the most common drugs, works something like this:

We know a drink will relax us.

It does this by first slowing down the part of the brain which *controls our behaviour*.

By the time we are cheerful, after two or three drinks, we have *lost this control*.

We are not able to make sensible decisions.

We should stop drinking and go on enjoying ourselves. But we say, 'I shouldn't really' as we hold out the glass for more, or, 'Well, just a quick one'.

Any further messages our brain sends out for us to stop are feeble and we ignore them.

By now we have totally *lost control*. We behave in silly ways, showing-off, crying, arguing, fighting, taking risks we wouldn't take if we were not drunk.

The drunker we get, the more we think it is a good idea to go on drinking.

As alcohol will affect us this way in one evening, you will understand how powerful it is. And how easy it is for people to become dependent on alcohol without knowing it is happening.

Fig 11.19

THE EFFECTS OF ALCOHOL

Alcohol always affects the body and the mind, even when you only have one drink. How much it affects you will depend on different things, like the rate of your metabolism and so on. Some people cannot drink at all as alcohol makes them sick and dizzy.

Alcohol does not have to be digested. It passes straight through the stomach wall and into the bloodstream. It is taken around the body, to the brain and all the tissues and organs. The only nutritional value it has is lots of Calories, so it is fattening and heat-producing. Drinkers sweat a lot to try to get rid of all the extra heat being produced. Some people, especially men, develop a huge swollen abdomen from too much beer drinking.

The first thing it affects is our *behaviour*. We lose our common sense, our self-control, our reasoning and our judgement. We are easily upset, laughing wildly, crying, fighting and quarrelling for no real reason.

Then it affects our *muscle control*. The alcohol interferes with the messages from the brain to the body and our movements are slowed down and clumsy. Words are slurred and repeated over and over again.

Then it affects our *senses*. We lose our sense of balance, dropping things and staggering slightly or swaying.

Speech, sight and hearing are affected. At this stage we are near to becoming unconscious.

There are plenty of jokes about drunk people. These make us laugh and we forget how serious the problem of alcohol really is. Can you imagine being a child in a home where one or both parents is an alcoholic? Or can you imagine having the person you most love being smashed to death by a drunken driver? People under the influence of alcohol can do terrible, cruel things, not only to themselves but to innocent people around them. If you have a heavy drinker in your family, you know only too well the misery and unhappiness and violence which goes on in a drunkard's home.

Alcoholism is a social problem as well. About half the people in prison committed crimes when they were drunk. The Accident and Emergency wards in hospitals fill up with drunks and their victims in the late evenings, after the public houses have closed. The wards for mentally ill people have their share of alcoholics. And thousands of people stay away from work each day because they drank too much the night before.

An alcoholic is a person who has no control over his or her drinking. They are now thought of as sick people who are in urgent need of treatment. In Britain there are special units in hospitals where anyone can go to be cured. From what you have read of the power of alcohol, you can guess that *no cure is easy*. But it can be done. There are many ways alcoholics can be helped to lead normal, healthy lives. A person who is cured can never drink alcohol again. Do remember, though, that prevention is always better than cure!

DRUGS

The word drug is used to describe any medicine taken to cure or prevent disease and to deaden pain. Drugs have been used for thousands of years. Chemical substances were taken from roots, seeds, leaves, bark and juices of plants. Sometimes they worked and the patient got better. And sometimes they didn't work. We still use the drugs that were successful, as well as the many powerful new drugs which have recently been discovered. Research chemists are now able to make drugs from different substances, as well as from plants. Treating an illness by the use of drugs is called **chemotherapy**.

foxglove
digitalis

Fig 11.20 Digitalis is an important drug in the treatment of heart disease. It has been used for hundreds of years.

THE SAFE DOSE

Discovering a new drug is a long and expensive business. It has to be tested and tried out on animals before it is used on people. A drug must be able to destroy the bacteria or viruses of a disease without damaging or destroying human tissue. If the drug is to work on the nervous system, it must be able to do so without damaging the delicate brain cells. This is very difficult and, unfortunately, nearly all drugs have some **side-effects**. Doctors have to work out very carefully the correct amount of a medicine so the patient will get better with the least possible side-effects. This is called the **'safe dose'**.

THE REASONS FOR DRUG CONTROL

1. Drugs affect different people in different ways. The level at which a particular drug harms rather then helps is different from one person to the next. Aspirin, for example, is thought of as a mild drug and is used to deaden pain or bring down a temperature. The safe dose for an adult has been worked out at two tablets. But some people, who take four times this amount or four doses a day, suffer from bleeding of the stomach and the intestines as a result. And larger doses of aspirin will be **fatal**, that is they will kill the person.
2. Some people build up a **tolerance** to a particular drug. Their bodies or their minds get used to the drug, so that each time they need it they have to take a larger dose.
3. Some bacteria and viruses may become **resistant** to a particular drug. This means the drug is not able to destroy them any more. When the person is ill the drug is useless against the germs and their toxins.

It is for these reasons that the taking of any drugs must always be carefully controlled.

Safety facts about using drugs

1. The prescription is made up for a safe dose for you.
2. It is unsafe to give your medicine or tablets to someone else.
3. Read the instructions carefully and take the exact dose.
4. Even when you are feeling well, finish all the drugs you have been given.
5. Don't take tablets or medicine in front of children.
6. Don't ever leave any drugs where children can get at them.
7. Throw away all unused drugs. They are not safe to take at some later time.

DRUG CONTROL

There are two ways in which we get drugs; from a prescription given by a doctor or by buying them from a chemist shop.

A prescription

This is a special form filled in by the doctor. It states the name of the drug, the amount of the drug, and the times it is to be taken each day. Before the doctor writes out the prescription, he decides what sort of illness you have. Then he works out the *safe dose for you*. He does this by checking your earlier illnesses, your height and weight, your body's metabolism and whether this particular drug will be a safe one for you to take. The prescription is then taken to the chemist who makes up the correct dose for you. The exact instructions for taking the drug are written on the label.

At the chemist

There are many drugs we can buy from a chemist without a prescription. Medicines for headaches, stomach upsets, coughs, colds and other less serious upsets are for sale over the counter. They are perfectly safe to take as long as you follow the instructions carefully. If you are not certain what you need, ask the chemist's advice. Most medicines bought over the counter in Britain do not cure illness. But they do take away the symptoms of an illness. For example, a stomach powder will remove the pain of indigestion but it will not cure you if you are suffering from an ulcer. Remember that the body is very clever at curing itself of small upsets if you have enough rest, exercise, proper diet and sleep!

Fig 11.21

Any other way of getting drugs is illegal. Most countries have very strict laws which forbid the buying or selling of drugs by people who are not licensed to do so. These laws are to protect us from the 'drug-pushers', the people who make huge fortunes by selling particular drugs which damage the brain cells, destroy the proper working of the body and cause early death. Buying, selling or taking illegal drugs is called **drug abuse**.

DRUG ABUSE

Only certain types of drugs are abused. These are usually the drugs which affect the working of the mind. They are for people who are mentally ill or under severe stress and for people who suffer great pain from some disease. The three main types are:

The pain killers

These include cocaine, morphine and heroin. They are very powerful drugs as they are able to deaden great pain.

The sedatives

These include barbiturates, which are strong sleeping-tablets and tranquillisers, which are anti-stress tablets.

The stimulants

These include amphetamines, which are 'pep' tablets, and hallucinogens, which are behaviour-changing drugs. Stimulants are not used by doctors nowadays, except in special cases, as they are so dangerous. (Marijuana, Pot, Hashish, Grass and Cannabis are a few of the names given to one type of drug which is not used by doctors at all. It is smoked, swallowed or sniffed by people who buy it illegally.) Not enough is known about the effects it has for a clear decision to be made about the damage it might do to the body or mind. It helps some people to feel happy and self-confident but a heavy dose can change the senses, so the person becomes frightened and upset.

WHO TAKES DRUGS ILLEGALLY?

Teenagers and young adults are the people most likely to take illegal drugs. There seem to be two reasons for this. The first reason is curiosity, the 'I'll try anything once' way of thinking. The second reason is that being a teenager is not always as easy as it is made out to be. It can be quite difficult at times, being neither a child nor an adult, being *between* ages. So he or she may take drugs to escape from problems and disappointments. All drugs change the way we feel. They affect large parts of the brain and the feelings have been described as 'having a high', 'leaving the world', 'going out of your mind', 'on a trip' and 'mind blowing'. When the effect of the drug wears off, the drug-taker suffers the *opposite* feelings. These can be so bad that he or she immediately wants more of the drug to stop the pain. You can understand how very quickly a person becomes dependent, both in body and mind, on drugs.

A great deal of growth and development of the mind and emotions happens during the years of being a teenager. Drug-taking slows down this growth; it can stop it completely. It can even turn the mind and emotions back into those of the child he or she was years ago. 'Mind blowing' sounds a very exciting thing to do. But think about those two words and think about the very complicated and delicately balanced working of the brain cells. And if you're mind is blown, how do you get it back together again – in the right order? Re-read the paragraphs on the safe dose and reasons for drug control. It has to be your decision as to whether you use or abuse drugs.

Fig 11.22

BUT OFFICER, DRINKING WITH CUSTOMERS IS PART OF MY JOB

Fig 11.23

Taking drugs is not a part of our social way of life in the way that alcohol is. And there are only a few drug addicts in comparison with the hundreds of thousands of heavy drinkers and alcoholics. But although the number is so small, we hear a great deal about the problem of drug abuse. This is because the effects of drug taking are so dramatic. The person quickly becomes dependent, then dangerously ill and, without treatment, he or she will die. It is also because the person most 'at risk' is the teenager, or the young adult who has not fully developed his or her mental controls.

Questions and things to do

1. Explain the meaning of the following words as simply as you can: receptor, effector, stimulus, series of impulses, synapse, neurone.
2. What is a reflex action? Copy the diagram (Fig. 11.7, page 138) and write your answer in your own words.
3. Explain the difference between a reflex action and a conditioned reflex action.
4. Imagine you are very hungry. You pick up your dinner plate which is so hot it burns your fingers. Instead of dropping it, you place it quickly on the table. Why do you do this? Make a list of all the messages, and the places they travelled to, which stopped you dropping the plate.
5. Describe the ways in which the brain and spinal cord are protected and get nourishment.
6. Why is it so important the brain has a rich supply of blood?
7. Copy out and learn the seven points explaining the function of the autonomic system.
8. Using diagrams, explain how the autonomic system works in an emergency.
9. In what way can parents help their baby to feel safe and secure?
10. What is meant by a mental handicap?
11. What do you think of Figure 11.23? Either have a class discussion or write your own story about it.
12. What is meant by a 'safe dose'?
13. What are the three main reasons for drug control?
14. Why is there no such thing as a safe dose of illegal drugs?
15. Copy out and learn the rules for 'Safety Facts on Using Drugs'.
16. Draw in the brain and spinal cord on the outline of a skater. Find out what happened before anaesthetics were discovered, and who first used them. Find out how Louis Pasteur discovered a treatment for rabies. Two American doctors, Dr Salk and Dr Sabin, discovered a way to fight polio (poliomyelitis). Read the history of the disease and the story of the doctors' work. Make a list of all the ways in which your local council helps mentally handicapped children and adults.

Chemical control

There is another system which also helps to control the proper working of the body, the **endocrine system**. It acts in a slower way than the nervous system as it sends its messages through the bloodstream.

The endocrine system

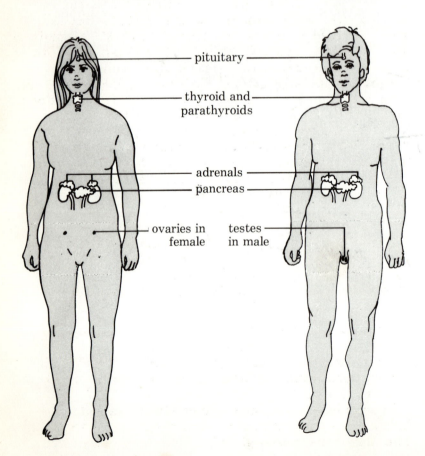

Fig 12.1 The endocrine system

The endocrine system is made up of **glands**. They produce chemicals called **hormones**. Hormones are powerful in their work. They control things like rate of growth and metabolism.

Endocrine glands do not have ducts. They are often called **'ductless glands'**. They secrete hormones straight into the bloodstream. Blood carries the hormones to where they are needed.

Some glands produce one or two hormones. Others produce two or more hormones. Some glands are controlled by other glands. So a fault in one gland may upset the working of another one.

Hormones have a powerful effect on the whole body. If a gland produces too much or too little of a particular hormone, we may get seriously ill. Without treatment we might die. There is a nerve centre in the brain which can 'tell' or 'judge' the level of hormones in the blood plasma. It is near the **pituitary gland**. It helps to control the work of the glands by sending messages to the pituitary.

The endocrine system is not under the control of our will. It is involuntary. Its function, with the nervous system and the autonomic system, is to control the proper working of the body. It is concerned with what happens *inside* the body.

The pituitary

This is a small gland, about the size of a cherry, just under the cerebrum in the brain. It produces many different hormones. Some of these control the work of other glands.

PRODUCTION OF HORMONES
The pituitary produces the **growth hormone**
This is a very important hormone, especially in children. It affects all the tissues and organs of the body. It makes sure we grow and develop at the right rate for us. Too much of the growth hormone makes a child grow too tall, which is called **giantism**; too little of the growth hormone and the child is too small, which is called **dwarfism**. Both these things are rare nowadays as a child's height and weight are regularly checked. Treatment is given if this hormone is under- or over-secreted. The normal difference in children's heights is due to the height of their parents. In adults, the growth hormone is used to repair and replace worn-out cells. Too much of the growth hormone makes the extremities grow; the hands, feet and jaw grow too large and too heavy. This is called **acromegaly**.
It produces the anti-diuretic **hormone**
This hormone is sent in the blood to the kidneys. It stimulates the kidneys to control the amount of water in the body. It does this by making sure enough water is re-absorbed back from the tubules and returned to the bloodstream. Having the right amount of water in the body is very important for the proper working of the cells, tissues and organs.

CONTROL BY THE PITUITARY
It regulates the working of the thyroid.
It regulates the working of the adrenals.
It stimulates and regulates the growth and working of the sex organs.
The pituitary controls, regulates, stimulates and balances the working of other parts of the body as well! It is often called the 'master gland'.

The thyroid

This is quite a large gland, wrapped around the front of the trachea just below the voice box. It uses iodine to function properly (page 75).

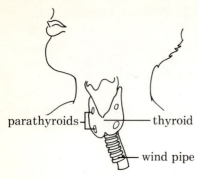

Fig 12.2 Diagram of the thyroid
and parathyroid glands

parathyroids — thyroid

— wind pipe

It produces the hormone **thyroxine**. Thyroxine controls the rate of our metabolism. This means it controls the rate at which food is used by the body; how quickly or how slowly energy is released and used by the cells and tissues. Too little thyroxine (under-secretion) slows down the rate at which energy is produced. Too much thyroxine (over-secretion) causes too much energy to be produced. Thyroxine is essential for the normal development of the body and the mind.

UNDER-SECRETION OF THYROXINE

This is very serious in a baby or child as neither his body nor his mind will develop. He becomes a **cretin**, with a short, thick body, an ugly face and an undeveloped mind. In adults, too little thyroxine causes **myxoedema**. The whole working of the body slows down. The person moves slowly, thinks slowly, speaks slowly. He loses his appetite but begins to get fat. The skin grows thick and coarse and the hair thin and dry. He doesn't have enough energy to live at the normal rate. Cretins are rare nowadays as thyroxine treatment is given at an early age so the child can develop normally. Adults with myxoedema are also treated with thyroxine to speed up their metabolism.

OVER-SECRETION OF THYROXINE

The hormone stimulates the body cells to work much too quickly. The heart rate and breathing increase and extra heat is produced. The person becomes thin, tired, and easily upset. He may eat a great deal but loses weight and gets thin. The thyroid gland swells, the eyes bulge out, and there is extra sweating to get rid of all the extra heat. This condition is treated by removing a part of the gland or by using drugs so that less thyroxine is made and the rate of metabolism is slowed down to normal.

The parathyroids

These are four tiny glands buried in the thyroid gland. They produce a hormone called **parathormone**. Its function is to regulate the calcium and phosphorous balance in the body. Too much parathormone causes calcium to be taken from the bones into the blood. This leaves the bones weak. The extra calcium in the blood is excreted in the urine. Too little parathormone causes lack of calcium in the blood. The person suffers from painful twitching of the muscles and a 'jumpy' nervous system.

Both the thyroid and the parathyroid produce other hormones as well.

The pancreas

You have seen the pancreas before and know it is an important digestive gland. It makes three different enzymes which collect in the pancreatic duct and pass into the small intestine (page 90). It is also an important endocrine gland, and this part of its work is quite separate

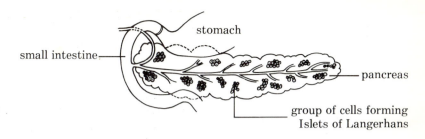

small intestine

stomach

pancreas

group of cells forming
Islets of Langerhans

Fig 12.3 The endocrine glands of
the pancreas

and has nothing to do with the digestive enzymes. Dotted
within the pancreas are groups of cells called **islets of
Langerhans**. They produce hormones which are passed
directly into the bloodstream.

INSULIN
The islets of Langerhans produce the hormone **insulin**.
Insulin controls the use of sugar in the body. It causes the
simple sugars in the bloodstream to be changed into stored
sugars in the liver and muscles. It also changes some into
fats. This keeps the right level of sugar in the blood. A
person who does not make enough insulin is called a
diabetic. Diabetics are not able to control the amount of
sugar in their blood. The level can rise far too high, or fall
far too low as sugar is lost by being excreted in the urine.
Diabetes can be treated by injections of insulin into the
bloodstream. Insulin is essential to life. Without it the
diabetic would go into a **coma**, become unconscious, and
die.

GLUCAGON
The islets of Langerhans also produce the hormone
glucagon. Glucagon has the opposite effect to insulin. It
changes the stored sugar back into glucose for use in the
tissues. Both insulin and glucagon have other important
functions.

The adrenals

These are two flat glands which lie one on top of each
kidney like small caps. They are nourished with a rich
blood supply.

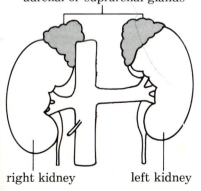

adrenal or suprarenal glands

right kidney   left kidney

Fig 12.4 The adrenal glands

THE OUTER REGION PRODUCES SEVERAL HORMONES
These hormones help to control the metabolism of protein,
starches and sugar, and common salt in the body. They
have many other important functions and help to control
the development of the sex organs.

THE INNER REGION PRODUCES THE HORMONE ADRENALINE
When we are in an emergency, or frightened, **adrenaline**
is secreted in much larger amounts than usual. It produces
the same action as the autonomic system; it prepares the
body for flight or fight (page 143). The autonomic system
sends fast messages along the nerves. The hormone
adrenaline acts as a 'back-up' system as it travels more

slowly in the bloodstream. It acts directly on the organs, blood vessels and tissues so its effects last much longer. Adrenaline is secreted any time our emotions are strongly affected. For example, an athlete waiting for the starting-gun is tense and 'keyed up' with excitement from adrenaline. This tension helps him to do extra well, maybe to break his own record. We all know those jittery feelings before an examination, speaking in public or going for a job interview.

The **testes** in the male produce a hormone, **testosterone**. The **ovaries** in the female produce two hormones, **oestrogen** and **progesterone**. We will study their functions in the next chapter.

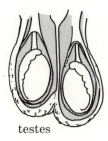

testes   an ovary

Fig 12.5 The sex glands

Health and hygiene of the endocrine system

As this system is not under the control of our will, there doesn't seem much we can do to keep it healthy and working well. Some forms of diabetes are linked with over-eating or being over-weight, so we can make sure we don't eat too much. We don't see many dwarfs, giants or cretins nowadays as treatment is given at an early age and so these sad conditions do not happen often. (Though there still are a very few parents who hide their baby away if they think something is going wrong with his development. These parents do not realize the baby's condition can be treated and cured.) All babies and small children must be taken for regular check-ups. Doctors and nurses are specially trained to notice when a gland is not working properly.

The endocrine glands are able to produce great changes in the body. Their importance has only recently been discovered. And more is being found out about hormones all the time. A hundred years ago a dwarf was thought of as 'one of Nature's freaks'. Now we know a dwarf is a normal person who did not have enough of the growth hormone (except in the case of a rare disease).

There is another interesting point about hormones. They help to make up the sort of person we are. When hormones are working properly, they still work at different rates for different people. This affects our behaviour. So, if you think again about the function of thyroxine or adrenaline, you

will understand just *one* of the reasons why we all have
different personalities.

**Questions and things
to do**

1. Find out and write the story of the discovery of insulin.
2. Make up your own table with the following headings:
   Gland   Position   Name of hormones   Function of
   hormones
3. 'The nervous system controls the proper working of the
   body.' Explain as clearly as you can how the endocrine
   system also helps control the proper working of the body.
4. How is an endocrine gland different from a digestive gland?
5. Write down a brief account of the main functions of the
   pituitary gland.
6. Write about the thyroid gland. What is its function and
   what happens if it produces too much or too little
   thyroxine? What treatment is given in each case?
7. Why is insulin so important in the body?
8. Give the names and positions of three endocrine glands
   and briefly describe the functions of each one.
9. Draw the endocrine glands, in their correct place, on one
   of the outlines of the skaters.

Chapter Thirteen

Passing on life

The main functions of the reproductive system are:
    To make male and female sex hormones.
    To make male and female sex cells, sperms and eggs.
    To bring the sex cells together so a new life is started.
    In the female, to provide a home for the unborn baby,
    and milk after it is born.
The organs which do this are the reproductive or **sex organs**. And the sex act, when the man and woman make love together, is called **sexual intercourse**. There are many other names for the sex organs and for sexual intercourse, but they are either too crude or too medical to use here. (You may like to have a class discussion about why some people use crude sex words when they swear or want to hurt some one else's feelings.)

Childhood

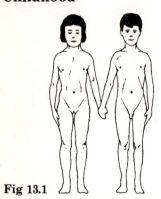

Fig 13.1

Puberty

The two children in Figure 13.1 look alike. If they were dressed in sweaters and jeans we might not be able to guess which was the boy and which the girl. Study the pictures carefully. The only difference we can see between them is their sex organs.

The boy has a penis and two testicles (balls). Inside his body there is a sperm tube and some more sex glands.

The girl has a slit where two folds of skin meet to form her vulva. Inside her body are the ovaries, the egg tubes, the womb and the vagina.

The reproductive system is not working in childhood. It doesn't begin to function till **puberty**.

Puberty is the time when a child's body changes and develops into the body of a young adult. It usually begins with a **growth spurt**. During our childhood we grew at a fairly steady rate. Just before puberty, we suddenly grew much more quickly. Then the body changes started. They usually start earlier for girls than for boys. Puberty begins between the ages of 11 to 13 for girls and between 13 to 15 for boys. But in some girls it can start as early as 9 or as late as 16. In some boys it can start as early as 11 or as late as 17. It doesn't matter if you start early, late or at the average time. It doesn't make you different in any way. It simply means you begin puberty at the right time for you. However, if not *one* of the things in the diagram has happened by the time you are 16, go to your doctor for a check-up.

Puberty begins because the pituitary gland in the brain begins to send out new hormones. These special hormones travel in the bloodstream till they get to the testicles in the boy and the ovaries in the girl. These are the **sex glands**. When the hormones from the pituitary reach them, they are stimulated to make sex hormones. The sex hormones made in the testicles and ovaries are sent into the bloodstream round the body. Their function is to change the body of the child into that of a young adult.

SECONDARY SEXUAL CHARACTERS

Fig 13.2 Secondary sexual characters

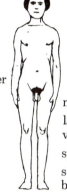

growth spurt

pituitary sends hormones

under arm hair

sex hormones made

sex organs get bigger

pubic hair

new sweat glands develop

layer of fat

breasts develop

hips widen

eggs are produced

menstruation begins

muscles develop

larynx enlarges so voice deepens

sperms are produced

some face and body hair

If you look back to Figure 13.1, you can already see the difference in the body shapes. You can get some idea of how very powerful sex hormones are. All these changes don't happen at once. Puberty takes between 2 and 3 years and a few secondary sexual characters, like facial hair on boys, may take much longer to develop. The importance of puberty is that the reproductive system begins to work. By the end of puberty the boy or girl is able to become a parent, to pass on life. It takes a longer time for the mind and emotions to develop. This happens later, during adolescence (page 180). Learn the **secondary sexual characters** now. They are the body changes which happen at puberty.

Structure and function of the female reproductive organs

The two ovaries
The ovaries are at the back of the abdomen below the kidneys. Each ovary is about the size and shape of an almond and is held in place by ligaments. The ovaries have two functions: to produce an **ovum**, which is the female

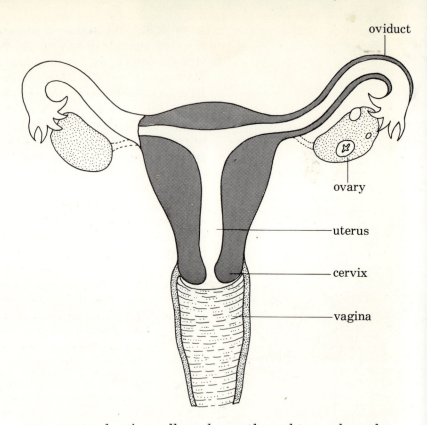

oviduct

ovary

uterus

cervix

vagina

Fig 13.3 Structure of the female reproductive organs

egg or reproductive cell, each month, and to produce the sex hormones, **oestrogen** (ee-stro-jen) and **progesterone** (pro-ges-ter-own). The ovaries begin to function at puberty and go on for the next 30 or 40 years. The time when they stop working is called the **menopause**, or 'change of life'.

The two oviducts

These are the egg tubes leading from the ovaries into the womb. They are about 10 centimetres long and very narrow. If you study Figure 13.3 you will notice the fringed ends of the ducts near each ovary. These ends draw in the egg as it leaves the ovary and sweep it into the oviduct. The oviducts have two functions: to pass the egg along to the womb by contracting their muscular walls, and to allow passage for the sperms to swim up and meet the egg so it can be fertilized in the oviduct.

The uterus

This is the proper name for the womb. It is about the size and shape of an upside-down pear. It has thick walls of muscle with a rich supply of blood vessels and glands. There is a soft inner lining which is shed each month if an ovum has not been fertilized. The uterus has two main functions: to provide a home and nourishment for the unborn baby, and to help in the birth by contracting the walls of muscles so as to push the baby out.

The cervix

At the lower end of the uterus is a ring of mucle called the

egg

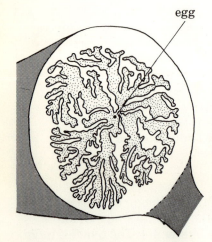

Fig 13.4 An egg in the oviduct

168

cervix. Its function is to close off the uterus so there is only a very small opening into the vagina.

The vagina

This is a tube made up of muscles leading from the uterus to the vulva. It is about 8 centimetres long and its walls are very folded. This makes it able to be stretched wide during the birth of the baby. It has special mucus glands to keep it moist and clean. It has two functions: to be the place for the man's penis during sexual intercourse, and to be the birth passage for the baby.

The vulva

Two folds of skin meet to cover the opening of the vagina and the urethra. Inside these folds is a small lump of tissue called the **clitoris**. It is very sensitive and gives the woman pleasure during sexual intercourse. Behind the clitoris is the tiny opening of the urethra. Behind the urethra is the opening to the vagina which, in some young girls, may be partly covered by a thin membrane called the **hymen**. The hymen was supposed to prove that a young bride was a virgin. Nowadays, we know the hymen can break at any time, while a girl is doing games, or dancing or taking exercise. This usually happens without the girl knowing.

The breasts

These are **mammary** or milk-producing **glands**. They grow out from the chest wall and are made of gland cells and fatty tissue. At the centre of each breast is a **nipple** with tiny openings for the milk. Around each nipple is a pigmented area called the **areola**. Milk is not made in the breasts till after a baby is born. It doesn't matter what size the breasts are as both large and small breasts can produce lots of milk.

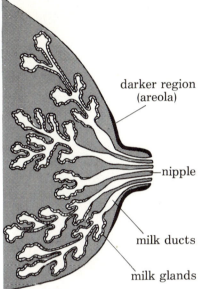

darker region (areola)

nipple

milk ducts

milk glands

Fig 13.5 Diagram of a section through a breast

OVULATION

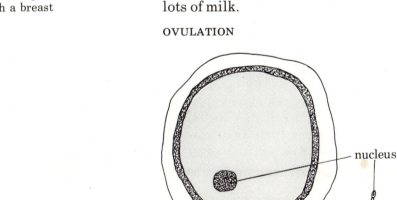

nucleus

Fig 13.6

egg

sperm

Each little baby girl is born with all the eggs she will ever produce already in her ovaries. When she gets to puberty, one of these eggs will ripen each month. The tiny hole it ripens in is called a **follicle**.

An egg is the female sex cell. It is a single cell with a large nucleus surrounded by cytoplasm. It is larger than any other cell in the body. It is about 2000 times larger than the male sex cell, the sperm. But it is only about the size of the full stop at the end of this sentence.

The egg ripens under the control of the sex hormones.
It bursts out of the follicle.
It is caught by the fringed ends of the oviduct.
It is passed down the oviduct and into the uterus.
It takes several days to travel along the oviduct.
If it is not fertilized, it is shed from the uterus during the monthly period.

THE MONTHLY PERIOD OR MENSTRUATION

Work out how many eggs a woman will produce from the age of 13 till she is 50, supposing she ovulates once a month and has no children. Then work out how many eggs she will produce if she has three children so doesn't ovulate for 12 months during and shortly after each of her three pregnancies. All these unwanted eggs have to be got rid of. But it isn't just an ovum which is removed each month. The lining of the uterus itself has to be shed as well. This is called **menstruation** (men-strew-ay-shun) or the **monthly period**.

At the beginning of each monthly cycle, just after the girl has finished her last period, hormones are sent to the uterus from the ovary. These hormones cause the lining of the uterus to become softer, thicker and richer. The blood vessels and glands swell up and get larger. A bird lines its nest with soft feathers before it lays its eggs: the uterus builds up its lining into a soft, rich 'home' or 'nest' to be ready for the egg which is travelling along the oviduct each month.

Halfway through the monthly cycle, ovulation happens and the egg begins its journey along the oviduct. The lining of the uterus is still getting thicker. Towards the end of the month, usually about a week after ovulation, the egg enters the uterus. If it is fertilized, it burrows into the rich lining and begins to develop into a new life. If it is not fertilized, no more hormones to make and keep the lining thick are sent. The blood vessels and glands contract and all the extra thickness which has been built up breaks away.

The carefully prepared nest, the lining and blood, mucus and the tiny egg pass through the cervix, down the vagina and out of the body. Menstruation lasts about 5 days. When it is over the hormones again build up the lining of the uterus as another egg is being ripened for ovulation and so, each month the cycle of building up and breaking down goes on. It is known as the **menstrual cycle**

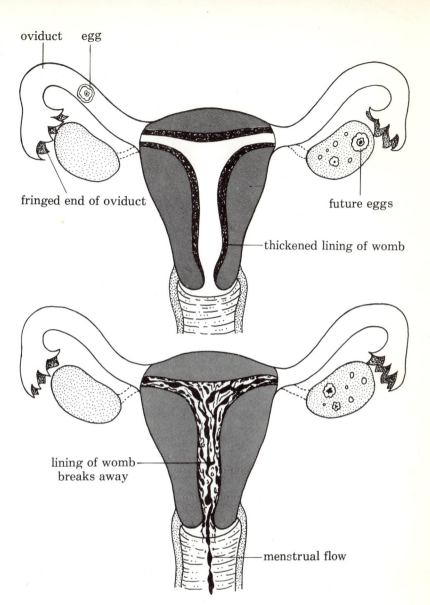

oviduct   egg

fringed end of oviduct

future eggs

thickened lining of womb

Fig 13.7 Halfway through the monthly cycle

lining of womb breaks away

menstrual flow

Fig 13.8 Menstruation

and is controlled by the action of both the oestrogen and progesterone hormones. The menstrual flow happens at the end of each monthly cycle but it is easier to understand the diagram on page 172 if it is put at the beginning.

Hormone control of the menstrual cycle is not easy to understand. All you need to remember is:
1. The pituitary sends stimulating hormones to the ovaries.
2. The ovaries make oestrogen and progesterone.
3. Both these hormones cause the lining of the uterus to thicken.
4. At the end of the month, the lining breaks down because only a little oestrogen and no progesterone is sent to the uterus.

Having a period affects different girls in different ways.

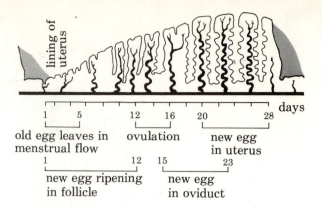

days

1   5   12   16   20   28

old egg leaves in   ovulation   new egg
menstrual flow   in uterus

1   12   15   23

new egg ripening   new egg
in follicle   in oviduct

Fig 13.9 The menstrual cycle

Most girls get used to it quite quickly. Some may get more spots or greasy hair or feel a little irritable just before the period begins. This is because of the changing levels of hormones in the body. A few unlucky girls get pain at the beginning of the period. They should make sure they are getting enough exercise, rest and sleep and are eating a proper diet.

We have studied the monthly period as if it happens every 28 days. This is only an 'average' and many girls have their periods every 21 days, or every 35 days, or any time in between. This does not mean you are *always* early or *always* late. It means that this is the right time for you. A 3 week or a 5 week cycle is perfectly normal. A few girls never have regular periods. When a woman wants to start a family, she should keep a note of the dates of her periods. When she becomes pregnant, this will help the doctor to work out when the baby is due to be born.

Remember that menstruation is controlled by hormones. So that things like extra excitement or nervousness or going on holiday can sometimes cause a period to be very late or to be missed altogether. However, if you do not have a period after you have had sexual intercourse, the usual reason is that you are pregnant.

Structure and function of the male reproductive organs

The two testes
The testes hang outside the boy's body, covered by a pouch of skin called the **scrotum**. The testes and scrotum together are called the **testicles** or balls. Each testis is about the size of a walnut and is made up of sex glands and sperm-producing tubes. The testes have two main functions: to produce millions of **sperms**, which are the male reproductive cells, and to produce the hormone **testosterone** (tes-toss-steer-own) and other male sex hormones. The testes begin to function at puberty and go on for the rest of the man's life.

The two sperm stores
Study Figure 13.10 to find out where they are. Their Greek name is **epididymis**, which means 'on the testes'. After the

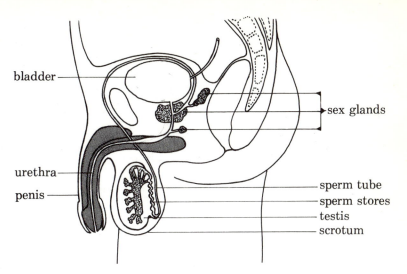

bladder

urethra

penis

sex glands

sperm tube
sperm stores
testis
scrotum

Fig 13.10 Side view of the structure of the male reproductive organs

sperms are made in the testes, they are sent to the sperm stores.

The two sperm tubes

These lead from the sperm stores up into the body. They are long because they have to loop around the bladder to arrive at the other sex glands. Their function is to carry the sperms to these glands.

The other sex glands

In Figure 13.10 you can see little glands and sacs near the bladder. They are called the **seminal sacs**, the **prostate gland** and **Cowper's glands**. They all produce fluids which are poured onto the sperms. The function of the fluids is to nourish the sperms and keep them active for the long journey to find the ovum. The mixture of fluids and sperms is called **semen**.

The urethra

This is a tube which leads from the sperm tubes and bladder down to the end of the penis. Its function is to be the passage for sperms *and* urine to pass out of the body. At the beginning of the urethra is a sphincter which closes to make sure urine and semen cannot pass down the urethra at the same time. The sphincter closes by reflex action so there is *no chance* that urine and semen get mixed up together.

The penis

The penis hangs outside the boy's body, in front of the testicles. It is limp as there is no bone in it. It is made up of spongy cells and blood vessels. The spaces in the spongy cells fill up with blood when the boy is sexually excited. The penis get large and firm so it stands out away from the body. This is called having an **erection**. The penis needs to be erect to get inside the vagina. The end of the penis is covered by a fold of skin, the **foreskin**. Underneath the foreskin is the most sensitive part of the penis which gives the man pleasure during sexual intercourse. The size of the penis is not important as it always gets large enough during an erection. The penis has two functions: to pass urine out of the body, and to pass sperms into the vagina of the female.

THE PRODUCTION OF SPERMS

Sperms are made in the testes. But a baby boy is not born with sperms as a baby girl is born with eggs. Sperms are not made till the boy reaches puberty. The testes of an unborn boy first develop *inside* his body. Before he is born, the testes move downwards till they get to the scrotum on the outside of his body. Sperms need a cooler temperature to develop which is the reason why the testicles are outside the body. A few boys are born with one or both testes still in the abdomen, **undescended**. This can be easily treated by a doctor.

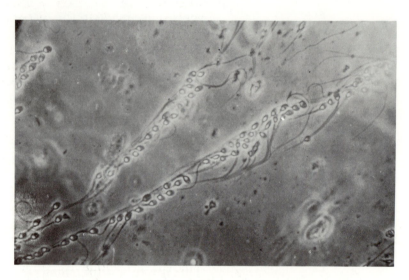

Fig 13.11 Human sperm

When puberty begins, the pituitary sends hormones to the testes to stimulate them to work. The testes produce testosterone and other hormones which change the boy's body into the body of a young man. The testes also begin to make sperms. Millions of sperms are made, and go on being made right throughout the man's life.

Sperms are the male sex cells. (The word 'sperm' is short for **spermatozoa**.) Each sperm is a single cell, looking rather like a tadpole. It has a nucleus in the head, and a tail which makes lashing movements so it can travel forwards. A single sperm is incredibly tiny. You can get some idea of its size by comparing it with the ovum, which is only the size of a full stop (Fig. 13.6).

Before you continue, make sure you know and understand the structure and function of the male and female reproductive organs.

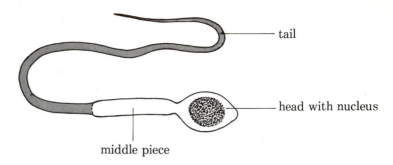

Fig 13.12 Diagram of a single sperm

tail

head with nucleus

middle piece

Health and hygiene of the reproductive system

CLEANLINESS

Urine, semen and secretions from the vagina are almost entirely free of germs. But faeces, the waste foods from the anus, contain millions of germs, dead and alive. As the anus is so close to the sexual organs, great care must be taken to keep the whole of the excretory and sexual area as clean as possible. At puberty, new sebaceous and sweat glands begin to function and pubic hair develops. The sweat and oils must be washed off daily with soap and warm water or they stick to the hair and skin and make a very good breeding place for germs.

BOYS

At birth, the tip of the penis is protected by the foreskin. By the age of 3, the foreskin can slide back easily, leaving the tip of the penis free. Underneath the foreskin are glands which produce a lubricating cream, **smegma**. Smegma helps the foreskin to slide back smoothly. All small boys must be taught to push back the foreskin gently and wash away the smegma with soap and warm water. If it is not removed, it becomes a breeding place for germs, causing serious infection and open sores. For a man, removal of smegma is even more important or he may risk infecting his partner during sexual intercourse.

In some cultures, it is the custom to **circumcise** small boys. The foreskin is pushed back, cut off, and the ends of skin are sewn together. You can see from the picture that the tip of the penis can now be easily cleaned. Parents have their sons circumcised from habit, for religious reasons, by custom or for hygiene. It is now believed the operation is unnecessary unless there is real difficulty in passing urine.

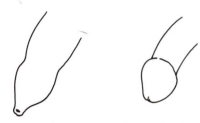

foreskin   circumcised

Fig 13.13 Uncircumcised and circumcised penis

GIRLS

There used to be a lot of 'old wives' tales' about the menstrual period. For example, if a girl touched meat while she was menstruating, the meat was supposed to be bad! Even today, a few silly ideas are still believed in. Menstruation is a perfectly healthy function and it is nonsense to think you should not bathe, wash or shower during a period. In fact, the opposite is true. A girl must take extra care over her hygiene during menstruation.

Many girls have a **discharge** from their vagina. These are normal secretions from tiny glands which keep the vagina moist, clean and healthy. When the discharge is fresh, it is usually creamy-white in colour and has no unpleasant smell. If it is not washed away daily, it mixes with the oils and sweat of the vulva. It forms a breeding place for germs and smells very nasty.

All children must be taught to keep their sexual organs clean as soon as they are old enough to wash themselves thoroughly.

Sexually-transmitted diseases

This is the correct name for the VD's, venereal diseases (Venus was the Goddess of Love) and for the other infections of the sexual organs. They are now called sexually-transmitted diseases because they are passed or transmitted from one person to another by sexual intercourse or sexual contact. Sexual contact includes petting, necking and touching the sex organs.

It is important to know that STD's are nearly always passed this way. They are *not* passed by dirty towels, baths, swimming pools or toilet seats. The tiny germs are quickly killed by the small amounts of chlorine used in swimming pools, by mild antiseptics and by washing with soap and warm water. Because they need moisture to stay alive, they die in fresh air, on toilet seats and towels.

We should be able to get rid of these diseases quite easily but, in fact, the opposite is true. There is now a huge rise in the number of people with STD's, especially young people. Gonorrhoea, for example, is now an epidemic disease, second only to the common cold! The reason why it is so difficult to control these diseases is that the tiny germs live *inside* the human body. Once a person is infected, the germs breed incredibly quickly. They travel up the mucus lining inside the penis or the vagina and spread right through the reproductive organs. They can also live inside the anus and rectum, and in the urethra in women.

Skin infections, such as warts, lice, scabies and sores, breed on the surface of the sex organs. Though they are not as dangerous as the germs which live inside the sex organs, there is the risk of secondary infection through the areas of broken skin.

STD's bring shame, misery, pain and embarrassment. It is difficult to understand why they are spreading so quickly. Nobody wants to get a disease of the sex organs. Nobody wants to be a disease-carrier. We all like to think of ourselves as healthy, clean and fresh and we want to think the same about our partner. How can we possibly help not getting any of these diseases as they are spreading so rapidly, especially among young people? There can only be one answer to this question. *Never* have casual sexual intercourse: *never* have sexual contact with a person you haven't known for some time. This is the only way you can protect yourself from catching a STD.

Look at the chart and study the lengths of time it takes the different diseases to appear. If you have casual sexual contact, your partner may not even know if he or she is infected. Neither will you know if you are already infected. You will know nothing about each other's standards of behaviour; of how careful, how clean or how responsible you both are about not passing on a STD. Think about these points:

The sore or sores which first appear on the sex organs from syphilis are not usually painful. After a while they heal up so the person may think he or she is better. But, in fact, the disease has passed into the bloodstream and the person is highly infectious. About a quarter of all women with syphilis do not know they are infected. This is because the sore grows inside the vagina and they are not aware of it; they cannot see it or feel it.

Many girls who are infected with gonorrhoea have no symptoms at all. The germs breed deep inside the sex organs. A woman may know nothing till she infects another person or finds she cannot have a baby.

Using a sheath gives some protection to both people.

TREATMENT

If you have had casual sex; if you have any reason to think you may be infected; if you are worried about any itchiness, sores, spots or a change in the discharge, *go to a STD clinic.* You can find out where your nearest clinic is by asking at the local Family Planning Centre. The doctors and nurses at the clinic will not make you feel dirty or guilty. They are there to help you get better. You will be treated in private and you can trust all the staff at the clinic to keep your secret.

The chart is to give you an outline of information about STD's. You must not use it to try to guess whether you, or any one you know, has a STD. Only highly trained technicians, after studying a sample of your blood and the lining cells of the sex organs, can tell whether or not a person has one of the serious diseases.

Remember:
1. The sooner a disease is treated, the easier it is to get rid of it.
2. The longer you leave it, the worse it gets.
3. You cannot treat yourself for the serious STD's.
4. Stop having sexual intercourse or sexual contact immediately.
5. Once you know you are infected, you must tell your partner.
6. Your partner may have no symptoms, but he or she must be treated.

SEXUALLY-TRANSMITTED DISEASES

Disease	Caused by	First symptoms	When first symptoms appear	What happens if disease is not treated	Treatment (Complete cure if treated early)
Syphilis (pox)	Bacteria Spirochaete	*Men*: Sore or ulcer on sex organs. Called a '*chancre*'. *Women*: Sore or ulcer on or inside sex organs. Sore is painless. Very rarely found in mouth.	Average 21 days. 10-90 days.	Bacteria enter blood stream and travel to all organs of body. In the late stage, it attacks the aorta, brain and spinal cord, causing heart disease, insanity, blindness, paralysis – death. (Foetus can get disease. **See page 212**)	Blood test shows syphilis bacteria. Course of injections. Check-ups for 2 years for complete cure.
Gonorrhoea (clap)	Bacteria Gonococci	*Men*: Burning pain when urinating. Smelly discharge or pus from penis. *Women*: 70% have no symptoms. Increase in discharge.	Average 3-5 days, not longer than 3 weeks.	The linings of the sex organs become swollen as bacteria breed. Scars form. The oviducts and sperm-carrying tubes become blocked. The person is sterile as eggs or sperm cannot pass through. (Baby can get disease at birth.)	Examination of discharge shows bacteria. Course of injections or tablets. Check-ups for 2 weeks after.
Non-specific urethritis (NSU)	Not yet known.	*Men*: Burning pain when urinating. Smelly discharge or pus from penis. A few men have only mild pain. *Women*: —	Average a few days to a few weeks.	Painful and long attacks of arthritis (disease of joints). Infections of the eyes, skin and mouth.	Different methods of treatment. May take a long time to treat.

Disease	Caused by	First symptoms	When first symptoms appear	What happens if disease is not treated	Treatment (Complete cure if treated early)
Trichomonases Vaginalis (TV) 	One-celled parasite. Protozoa	Men:— Women: Yellow, smelly discharge. Pain and itchiness. Soreness of the vagina and vulva. (Urination maybe painful).	Less than a month.	Lasting vaginal discharge, pain and itchiness.	A week's course of tablets.
Candidiosis (Thrush) 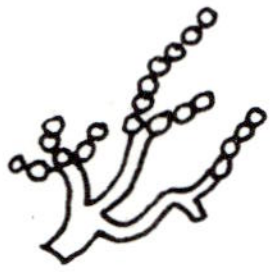	Fungus	Men: Quite rare in men. Tip of penis red and sore. Women: Thick discharge. Soreness and itchiness of vagina. (Not always an STD).	A few days to a few weeks.	Chronic vaginal discharge, pain and itchiness.	Two weeks' course of tablets, ointments, or pessaries (tablets put inside vagina).
Warts 	Virus	Looks like tiny cauliflower on penis or vulva.	A few days.	Risk of secondary infection.	Special creams and ointments. Strict personal hygiene.
Herpes 	Virus	Look like 'cold sores' on penis or vulva.	A few days.	Risk of secondary infection.	Strict personal hygiene.
Scabies, Lice (Crabs) 	Skin parasites	Itchiness. Can be seen on pubic hair.	A few days.	Risk of secondary infection. Parasite spreads to rest of body.	Treatment with DDT lotion. Strict personal hygiene.

Fig 13.14 Sexually transmitted diseases

7. You are not *immune* (page 269) after an attack. You can be just as easily re-infected.
8. Obey the instructions from the clinic exactly. Always keep your appointment. They will tell you when you are cured.

Adolescence

Developing from a boy or a girl into a man or a woman isn't just a physical thing which happens. It isn't just a matter of sex hormones and sex organs. It is far greater and deeper and more important. It is concerned with the whole growth of your personality. Adolescence is the time of great mental and emotional and social development. Our ways of thinking and feeling and behaving change as we learn to understand more about the adult world we are moving towards. It takes a long time to change from a child to an adult. So, though adolescence begins at puberty, it goes on for much longer, till we are 19 or 20. For some people, it goes on even longer.

Here are a few points you might like to think about, or discuss in class.

CULTURAL VARIATION

We are all conditioned by our families, our schools, our friends, our society, our religions, our laws, to think and feel and behave along certain pathways. We learn our moral values, what is 'right' or 'wrong' behaviour, from our **culture**. This is especially true of sexual behaviour.

Fig 13.15 Is it any more odd to pierce your nose than it is to pierce your ears?

Each culture has its own laws and customs about what is right or wrong. They are made to help and protect the people in that culture. What may seem silly or wrong to the people of one culture is, in fact, sensible and right for the people of another culture. During adolescence, it is useful to examine our own laws and customs so we get a better understanding of our society and of human behaviour.

Fig 13.16 a We all develop at different rates
  b This chart shows average heights and variation

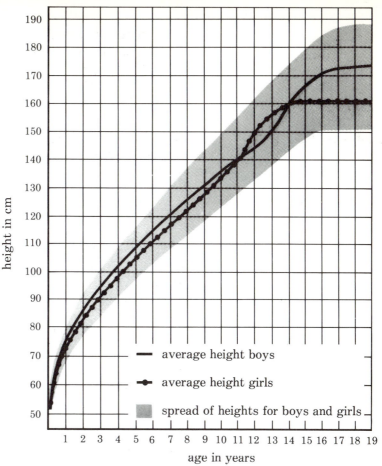

height in cm

— average height boys

—●— average height girls

▓ spread of heights for boys and girls

age in years

HUMAN VARIATION

Each human being is *unique;* each person is an *individual;* there is no one person quite like anyone else in the world. No one person will think or feel or behave in exactly the same way as any other person. This is called **human variation** and life is much richer and more exciting because we are all different. During adolescence we learn to accept that people *are* different. We learn to understand that being different doesn't make others 'better' or 'worse' than we are.

Another point about human variation is that our appearance, the way we look, our figure and face, and our rate of development, are different from anyone else's. It is important to remember this, or you may spend too much time worrying about how different you think you are. Adolescence is the time for mental, emotional and social development. Don't waste it worrying about your growth. (You can't control your hormones, anyway!) Use the time for the growth and development of your own personality.

THE SEX DRIVE

This is the natural, normal drive to reproduce ourselves; to pass on life; to have children of our own.

A boy's sex drive starts quite early in puberty when his testes begin to make millions of sperms. As the numbers of sperms increase, he feels sexual tension and the urge to get rid of them from his body. This causes him to have erections; the first ones usually happen at night. He wakes, in the middle of an exciting, sexual dream, to find he has had an **ejaculation**. This means the semen has passed out of his penis, and is more often known as 'coming' or 'coming off'. Having an ejaculation at night is called a 'wet dream' and the boy has no control over it. He may also get erections during the day, which can be embarrassing for him. When he is alone, he may **masturbate**, rub the penis by hand, so that he does not have unwanted erections. He feels pleasure and a release from sexual tension.

A girl's sex drive begins more slowly. She does not have the same experiences as a boy as she has no sperms to get rid of. But a few girls do feel sexual tension and the need to masturbate; they rub the vulva by hand. Masturbation is not harmful, nor is it 'wrong'. It is just a method some people use to get rid of sexual tension before they are married.

Our sex drive begins in adolescence; before we are ready to get married, to build up a home, to have children. This may cause a few problems for some people. (Remember, we all think and feel and behave in different ways.) We may get quite strong feelings of wanting to be adult and have sexual intercourse while we get equally strong feelings of wanting to wait, to take things slowly, to have more time to sort out what we really feel and what we really need.

Flirting, kissing and 'petting' or 'necking' are all ways of trying out new feelings. Because these *are* new feelings, you haven't yet learned how to control them. These things don't happen to us any more. We have had many years to learn what we can and can't do. We need to take plenty of time to learn to cope with and to control our sexual feelings. We have to learn not to 'fall down', not to make mistakes. The children in Figure 13.17 can easily have their mistakes sorted out. They are not serious mistakes. But sexual mistakes are not easily sorted out. And they are nearly always serious. We risk becoming a parent before we are ready. We risk catching a sexually-transmitted disease. As well as these serious risks, we may damage ourselves emotionally. Many adolescents suffer fear, anxiety and great heartache from making sexual mistakes. Without meaning to, they also damage their partners' feelings.

Fig 13.17 These things don't happen to us anymore

Love and marriage

Towards the end of adolescence we begin to want a steady relationship with one other person. We start to search for the one person we can love and trust completely, and who

Fig 13.18

will love and trust us completely in return. Finding exactly the right person takes a long time. We may fall in and out of love quite often before we find the one person we want to share the rest of our lives with. While we are searching, we need to be very careful we don't mistake sexual desire for love. Some young people get married because they believe their strong sexual feelings for each other mean they must be in love. Sadly, after their marriage, they realize they are not in love at all.

Growing up to love and marriage can be tremendously exciting. It is an enormous step forward in your development as a person. Finding out your real needs (not just the things you think you want) can be quite difficult. Learning about your partner's needs is equally important. Remember, you will need to suit each other mentally, emotionally and socially – as well as physically. Remember also, because of human variation, there will be certain differences between you. Love which is deep and true helps to smooth over the differences and will make your marriage happy, loving, rich and rewarding.

Questions and things to do

1. What is meant by puberty? What causes it to begin?
2. Write out the list of all the things which happen (*a*) to a girl and (*b*) to a boy, at puberty.
3. The testes and the ovaries are the male and female sex glands. Clearly explain their functions.
4. Draw and describe the male and female reproductive cells.
5. What is menstruation? Answer this question as fully as you can.
6. Learn and be able to draw from memory the male and female sex organs.
7. With the help of a diagram, explain the passage of an egg from ovulation to menstruation.
8. With the aid of a diagram, show clearly the passage of a sperm from the testes to ejaculation.
9. In what ways will an understanding of cultural and human variation help adolescents to develop more fully?
10. Love and marriage is not suitable for every person. Have a class discussion on why you think this might be true.

Chapter Fourteen

A new life begins

Fertilization

When a man and a woman have sexual intercourse, the
sperms are ejaculated from the man's penis far up inside
the woman's vagina. If you look again at the diagram of the
female sex organs (Fig. 13.3), you will see how far the
sperms still have to go before they reach the egg.

They have to pass through the tiny opening of the
cervix.
They have to travel right up through the uterus.
They have to swim into the oviduct.
They have to travel along it till they find the egg.
This is an incredibly long journey for them. The fluids of
the semen help to keep them active and nourished. From
the millions of sperms produced at each ejaculation, only
a small number will finally get to the egg. It seems
reasonable to think that the fastest and strongest sperms
get to the egg.

Sperms can stay alive for a few days in the female sex
organs, but the exact length of time isn't yet known. If
there is no egg in the oviduct, the sperms will die. When
there is an egg in the oviduct, the sperms swarm around it,
trying to get in. The ovum is protected by an outside
envelope, the **corona**. The sperms press against the corona
until it gives way.

But only one sperm is allowed inside the egg. It loses its
tail and its head swells up as it travels to the heart of the
ovum, the nucleus. The nucleus of the sperm and the
nucleus of the ovum **fuse** together. This is the moment of
fertilization. The first cell of a new human being is formed.
The fused nuclei carry all the instructions needed to form
a new life. Fertilization is also called **conception**.

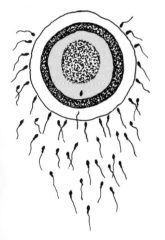

Fig. 14.1 Only one sperm is
allowed inside the egg

Implantation

After a while, the fertilized egg divides. First into two
similar cells, then four, eight, 16, and so on, till there is a
tightly-packed ball of cells. The walls of the oviduct push
the ball of cells along to the uterus. A few days after
fertilization it enters the uterus, where the thick rich
lining is waiting for it. The egg burrows into the lining –
and the lining moves upwards to surround the egg. This is

185

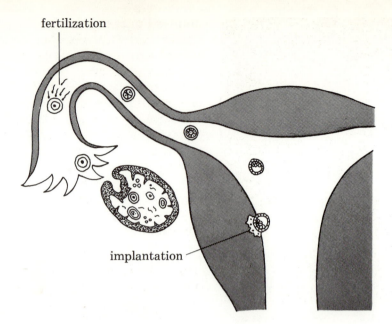

fertilization

implantation

Fig 14.2 How the fertilized egg implants itself in the womb

called **implantation**; the egg and the lining both helping to 'plant' the new life into the uterus. Messages are sent to the ovaries to continue sending progesterone so the lining stays thick. There must be no menstruation now until after the baby is born.

The embryo

During the first weeks after implantation, the dividing cells are called the **embryo**. It is now believed that the embryo stage of a new baby's life in the womb is the most important of all.

THE FIRST MONTH

The embryo doesn't yet look much like a human baby. But it doesn't look much like the ball of cells either. It is now 10 000 times larger than when it implanted, though it is only 5 millimetres long! Find 5 millimetres on your ruler. Cell specialization has already happened. Each cell that is formed carries its instruction; to be a muscle cell, a nerve cell, a skin cell, and so on. Each cell moves to its right place in the embryo to begin to build up the head, the body and the limbs. The outline of the face is already being formed. There are little dimples and dips for the nose, the eyes, the ears and the mouth. The limbs are beginning as tiny buds growing out from the embryo. It is all quite magical and mysterious, as silently and swiftly the cells and tissues form themselves into human shape.

THE SECOND MONTH

The cells and tissues are now forming into nerves and organs; lungs, liver, kidneys, bladder, stomach, intestines, brain. The tiny heart has been beating for almost a month (and will go on beating till death). Footprints and

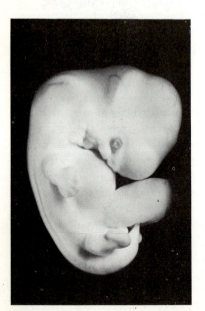

Fig 14.3 A 4 week old embryo

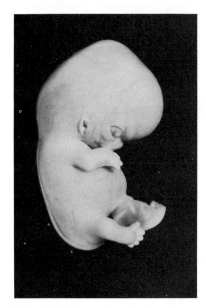

Fig 14.4 This embryo is 8 weeks old. Notice how much it has developed

palmprints can be seen as tiny lines on the hands and feet and the nails are even beginning to grow. By the end of the second month, the soft cartilage starts to ossify, to harden and begin to form into bones. (You will remember that ossification is not finished till early adulthood.) The tiny embryo is now 3 centimetres long and weighs only about 1 gram. Use your scales to find out other things weighing 1 gram (e.g. an envelope). It is most important to remember that during these first 2 months after fertilization, the embryo has developed the outlines and beginnings of *all it will need* to grow into a human baby. For the next 7 months, growth and development continue of the organs, tissues and cells the embryo has formed in the first 2 months.

CARE OF THE EMBRYO

Some years ago, certain babies were born in certain countries with terrible, tragic deformities. Their limbs, parts of their skeleton and organs were either missing, stunted or not fully developed. Through careful tracing of the facts, it was found that all the mothers had taken the drug Thalidomide during the early months of pregnancy. You might like to find out more about this tragedy and what society is doing to help these children who are now almost grown up.

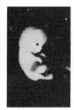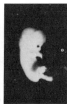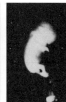

Fig 14.5 Parts of a ciné film showing a 7 week old embryo moving in the womb

But from this tragedy, an extremely important and useful lesson was learned. It wasn't just the lesson that pregnant women should not take drugs. It was also that it is during these early months of development the embryo is most at risk, most likely to be harmed. The whole structure of the new baby is being formed in the first months.

But the mother of the embryo in Figure 14.3 has only missed one period, one menstruation. She may just think she is late. She may decide to wait till she has missed her next period before finding out if she is pregnant or not. She may have no idea there is a tiny embryo developing so swiftly inside her. How can she make sure she doesn't harm her baby if she doesn't even know she is pregnant?

Parents who plan their family decide when they would like to start a baby. We cannot tell when one sperm will fuse with an ovum, so the mother starts to prepare herself *before* she wants to get pregnant. Her body, the baby's first environment, will be strong and healthy and ready for child-bearing.

BEFORE BECOMING PREGNANT

The mother should have the condition of her teeth checked. The foetus will need plenty of calcium and phosphate for the formation of its bones. The dentist can advise her about her diet.

If she is having treatment for an illness, she should tell her doctor she is planning to start a family. The doctor can then discuss details of her health and her future pregnancy.

If she smokes cigarettes, she must break herself of the habit. Smoking can, and does, harm the foetus.

She needs to make sure she has had the injection against rubella as the virus can pass through the placenta and damage the foetus.

DURING PREGNANCY

If she needs any X-rays, she must tell the radiographer she is expecting a baby. Then special precautions will be taken so the X-rays do not damage the foetus.

The only drugs she should take are those prescribed by her doctor. She must not take any others.

She needs to have a well-balanced diet with plenty of protein foods for cell formation and body building.

She should take sensible exercise each day and later in her pregnancy, she may want a rest when she gets tired. She can carry on with her usual routines. Most women are healthy and active during pregnancy.

She must visit the doctor and/or ante-natal clinic regularly so that proper checks can be kept on her health and that of her developing baby.

The foetus

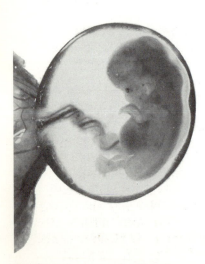

Fig 14.6 Seven week old embryo in its amniotic sac

For the remaining 7 months of pregnancy the unborn baby is called a **foetus**. During this time, the cells, tissues and organs will continue to develop and begin to function. The foetus will grow from 3 centimetres to 50 centimetres and its weight will increase from 1 gram to about 3 kilograms. Towards birth it stops looking so wrinkled as it grows a layer of fat for warmth and as a food store.

We imagine that life in the uterus is very pleasant. You can see from Figure 14.6 that the foetus is floating in a bag of waters, the **amniotic sac**, which keeps it warm, weightless and protected from knocks and bumps. It moves, yawns and stretches, kicking out for exercise and turning over to a more comfortable position. It falls asleep and wakes up. It can hear very loud noises and may kick out if it doesn't like them! It sucks a thumb and may swallow a little of the amniotic fluid. It may grow hair and be covered in a creamy grease which protects the skin from the waters. From the fifth month the heart beats can be heard through a **stethoscope**. Though it cannot breathe air into its lungs, nor take food into its mouth, it gets all the oxygen and nourishment needed from its mother.

The placenta and the umbilical cord

These are two special reproductive organs which only grow in pregnancy. Once the baby is born, they are got rid of as the 'afterbirth'. Each time a woman is pregnant, she must grow a new placenta and cord for the new baby.

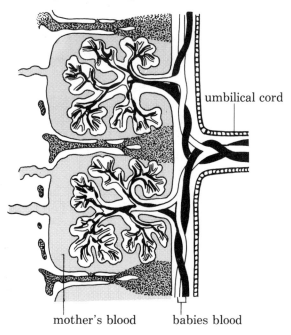

umbilical cord

mother's blood   babies blood

Fig 14.7 How things pass through the placenta

THE PLACENTA

The name comes from a Greek word meaning a 'flat cake'. It grows from the embryo tissue and the tissue on the lining of the uterus. It is soft and spongy and very folded as it is full of villi with their tiny blood vessels. As the foetus gets larger, so does the placenta. At birth, the placenta will be about 20 centimetres across, weigh about 450 grams and will look rather like a flat cake, though a very red-coloured flat cake.

The placenta is the organ which passes dissolved oxygen and foods to the developing foetus. It also takes back the waste products made by the foetus during metabolism. In the placenta are tiny blood vessels from the mother and tiny blood vessels from the foetus. *They are separated by very fine membranes.* The mother's blood carries all the things needed for the development of the foetus. (You should be able to make a list of all the things carried in the blood.) These substances pass out of the mother's blood vessels, across the fine membranes, and into the blood vessels of the foetus. The mother's blood does not pass across the membranes. Nor does the foetus's blood pass across the membranes. The blood of the mother and foetus do not mix. They have separate bloodstreams.

The placenta acts as a barrier by stopping harmful things in the mother's blood from passing through. Unfortunately, there are some damaging things it is not able to filter out. The viruses of some diseases can pass through and cause serious harm to the foetus, like the Rubella virus. Drugs can pass through; so all pregnant mothers are warned not to take any drugs, even aspirin, *unless* they are given to her by her doctor. Smoking is also a danger; babies are born lighter and less fully developed to mothers who smoke. They may have trouble with breathing and have other health problems.

The diet of a pregnant woman is very important. She doesn't need to eat for two people, but she must make sure she has *enough of the right foods*. Vitamins and minerals, especially iron, calcium and phosphorous, are vital for the healthy development of the foetus. If they are lacking in her diet, her own body stores are 'robbed' to give to the foetus. The old saying 'A tooth lost for each baby' means that because the mother was lacking calcium and phosphorus in her diet, they were taken from her own stores. She may well have lost a tooth! Nowadays, all caring mothers go to their **ante-natal** (before birth) **clinics** right through their pregnancies. They are given diet sheets, extra vitamin or iron tablets if necessary, and everything is done to look after the health of both mother and her developing baby.

By the end of the third month of pregnancy, the placenta is producing its own hormones to make sure menstruation doesn't happen and the foetus will continue to develop.

Fig 14.8

THE UMBILICAL CORD

The cord develops, with the placenta, in the first months of embryo life. It grows out from the tiny blood vessels of the placenta and into the abdomen wall of the embryo. As the foetus gets larger, so does the cord till, at birth, it is an average of 50 centimetres long. This allows the unborn baby plenty of room to move around without getting the cord twisted.

From Figure 14.7 you can see that inside the protective covering there are two arteries and one vein which carry the blood from the foetus to the placenta and back again. The umbilical cord is the vital link between the mother and developing baby.

THE FOETAL CIRCULATION

Blood, rich with dissolved oxygen and foods, leaves the tiny blood vessels in the placenta. It travels along the vein in the cord and enters the foetal abdominal wall. It is pumped by the foetal heart to all the cells of its body, giving off oxygen and foods and picking up waste carbon dioxide and the waste products of metabolism. The blood is then pumped back along the two arteries of the cord and into the tiny blood vessels in the placenta. Its waste products are passed through the membranes into the mother's blood vessels and more oxygen and foods are picked up. The cleaned and enriched blood once more travels back to the foetus. The waste products are taken back into the mother's circulation and are removed along with her own excretions.

The birth date

The day of birth is worked out as 280 days, or 40 weeks, from the start of the woman's last menstruation. Of course, many babies are born days earlier or later! The correct name for the length of a pregnancy is the **gestation time**. When you plan your baby, it helps if you keep a definite record of your periods so your doctor can work out a more accurate gestation time.

A **premature** baby is born before the gestation time has finished. The baby weighs less than $2\frac{1}{2}$ kilograms and is very thin as there has not been time to grow the layer of fat needed for protection, for warmth and as a food store. Though all the organs may be fully developed, the baby may have breathing troubles, may not be able to feed, and may suffer from lack of body heat. Premature babies need special nursing care in a really warm **incubator** if they are to survive.

The birth

During the last 6 to 8 weeks of gestation, the foetus will turn so that its head is downwards, towards the cervix. This is the correct position for it to be born; only 3 per cent of babies are born feet or bottom first. The uterus is now fully stretched and the foetus, which is now **full-term**, is ready to be born. Birth happens in three stages.

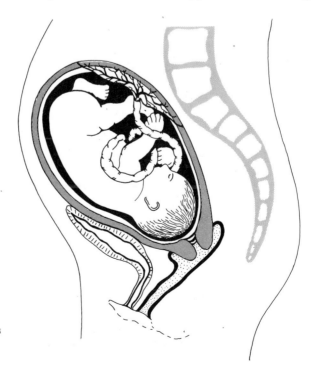

Fig 14.9 A foetus just before it is born

THE FIRST STAGE

The mother feels twinges across the back and front of her abdomen. These twinges are the very first **contractions**; the first squeezing of the muscles of the uterus. They come quite regularly, at first every half an hour, then every 20 minutes, every 10 minutes, and so on. Usually they are not painful at first. The mother can tidy herself, pack her bag and pass the waiting time reading, chatting, playing cards. It is important for mothers to understand this first stage of birth is the longest. It takes about 14 hours and this can seem a long time to keep calm, and cheerful and patient. By the time the contractions are coming every 5 minutes, the mother is ready to go to hospital. In Western countries, nearly all first babies are born in hospital.

As the muscles of the uterus contract, the foetus is pressed down onto the cervix. Under this constant pressure, the cervix muscles slowly begin to open. They must open wide enough for the baby's head to be able to pass through. The amniotic sac usually breaks during the first stage and the 'waters', the amniotic fluid flow out of the vagina.

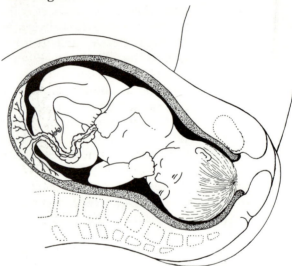

Fig 14.10 The cervix is opening slowly

Having a baby is called **labour**, because it can be very hard work indeed. By now, the contractions are very strong and some women, though not all, do feel quite a bit of pain. There is no need to worry about having a painful labour as the **midwife** and the doctor will give pain-killers to the mother. However, any sort of drug will affect the baby as well, so it is not wise to ask for drugs till you absolutely need them. Mothers who have been to the ante-natal classes know all the special exercises and breathing methods to help them have an easier labour.

THE SECOND STAGE

This is the actual birth of the baby and happens quite

quickly, lasting for an hour or so. The cervix is now completely opened. The mother feels very powerful pushing contractions. As the baby enters the birth canal, the vagina, the mother helps by pushing or 'bearing down'. At last, the baby's head can be seen and with a few more pushes, the head is born. The shoulders and the rest of the body quickly follow. A new baby is born. A new life has begun.

THE THIRD STAGE

The muscles of the uterus stop contracting for a short while after the baby is born. The placenta comes away from the wall of the uterus. Then the muscles contract again, so the placenta, the protective linings and the cord are pushed out of the uterus. These waste products of pregnancy are called the **afterbirth**. They are carefully examined to see that nothing has been left behind.

BIRTH FOR THE BABY

It has been a long and difficult journey for the baby. He has been pushed out from his snug home into unknown space. The cord, his life-line, has been clamped and then cut. He has taken his first breath of free air. He has shouted his first protests. He has been weighed, measured and checked for normal development. He needs comfort, sleep, peace and safety. He needs his mother.

Breast feeding

During pregnancy, the mother's breasts get larger as the glands and ducts develop. After the birth of the baby, when the placenta is removed, a hormone, **prolactin**, is produced in the pituitary. Prolactin acts on the breasts and starts milk production. At first, a yellow liquid called colostrum is produced. By the third or fourth day, the mother's milk is flowing easily. The sucking action of the baby causes more prolactin to be produced from the pituitary, which causes the breasts to make more milk.

Human milk is the best food for human babies. It has now been proved that dried, tinned or powdered cows' milk can never match the quality of human milk. Breast milk has *exactly* the right *balance* of proteins, fats, carbohydrates, vitamins, mineral salts and water. And the amounts of these things will change and adapt to suit the baby. Babies fed on breast milk get less diarrhoea, less bronchitis, less coughs and colds, less digestive upsets, less nappy rash. They get protection from certain diseases because of the antibodies in the mother's milk. They put on weight without putting on extra fat. Breast feeding helps to form the close, deep and loving relationship between a mother and baby. All premature babies are fed on breast milk as part of the extra special nursing care.

It is difficult to understand how bottle-feeding became so popular. You might like to have a class discussion on

Fig 14.11

193

this now. In underdeveloped countries, among mothers who want to follow our fashion for bottle-feeding, many babies die because the artificial feeds are not made up properly and because it is difficult to sterilize the bottles, teats, mixing jugs and spoons. Bottle-feeding was thought to be a step forward in progress. For the baby, no matter what country it lives in, it is a step backwards.

Twins

Twins are born once in every 87 to 100 births, Sextuplets are born once in every 5 000 000 000 births. You will be more likely to have twins if there is a history of twins in your's or your partner's family. There are two types of twins: non-alike or **fraternal twins**, and similar or **identical twins**. Nearly three-quarters of all twins are fraternal.

FRATERNAL TWINS
These are the result of *two* different eggs being fertilized by *two* different sperms. This happens when the mother, for some reason, sheds two eggs in one month from her ovaries. The two eggs are then fertilized by two different sperms, travel to the uterus and become implanted. They develop two separate placentas. The two babies grow in the womb and are usually born within a short time of each other. Fraternal twins can either be the same sex or be one boy and one girl. They are no more like each other than any brothers and sisters in one family.

IDENTICAL TWINS
These are the result of *one* egg being fertilized by *one* sperm and then, for some reason which is not understood, the fertilized egg splitting into two identical eggs. These two identical eggs then develop into separate embryos but they do not have separate placentas. You can see from Fig. 14.12 that identical twins share the same placenta. Identical twins will always be the same sex. Can you work out why this is so?

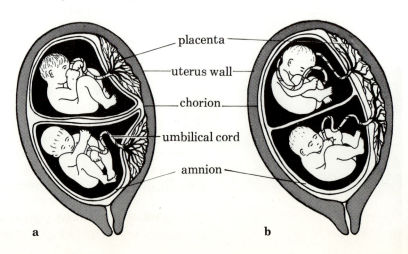

placenta
uterus wall
chorion
umbilical cord
amnion

Fig 14.12 a Fraternal twins
b Identical twins

a   b

Infertility

Sadly, there are some couples who are not able to have babies. The man may not produce enough fertile sperm. The woman may not produce eggs or her oviducts may be blocked. There may be other reasons for infertility. In many cases the couple can be treated at a fertility clinic and be able to have the baby they want so much. (You may have read of 'multiple births'. These happen when the woman is given hormone treatment to help her produce eggs, and she produces too many in one month.)

The point to understand about infertility is that the man is no less masculine, nor is the woman less feminine, because there is a flaw in one of their reproductive organs. They are able to have a happy and full sexual life. If treatment does not help them, they can adopt a baby if they want to. It is wrong and unfair to think of people as 'odd' or 'not normal' if they can not have a family.

Family planning

Family Planning, Birth Control, Contraception and Safe Sex all mean the same thing. They mean we can now control our own fertility. We can decide when we are ready to start our family. We can choose when we want to have a baby. We can decide and choose *when* and *if* we want to be parents by controlling our fertility. Babies no longer have to be 'mistakes', 'unwanted', 'accidents'. It is a marvellous thing we now have contraception. There is hope that babies of the future will be born into happy and loving homes; where they are wanted, cared for and loved.

We have learned that our sex drive begins while we are adolescents; before we are mentally and emotionally developed; long before we are mature enough to be successful parents. Of course, people must make their own decisions about their sexual life and these are personal, private decisions. But when the results of the decisions are unwanted babies, or families started before the couple can support them, then it becomes a matter of public and social concern.

There is now a great amount of evidence that babies have a chance of better development and of a happier and healthier life if:
a. the parents are not too young;
b. the parents are happily married;
c. the parents are mature physically, mentally, socially emotionally;
d. the parents can understand they must put the baby's needs before their own needs.

There are many, many arguments for and against contraception. If you have a class discussion on the subject now, you will learn some of the different opinions about it.

Abortion

This is the correct medical name for a miscarriage. There are many different reasons why a developing embryo dies and is passed out of the woman's body. Miscarriages usually happen in the first 3 months and, though it is sad for the parents-to-be, one of the main reasons is there is something wrong with the embryo.

However, abortion has another meaning, which is the removal of an unwanted embryo or foetus from the mother's uterus. You will see from the chart that not all methods of contraception are safe and, no matter how careful the couple are, unwanted babies are still conceived.

Whether or not abortion is right or wrong, should be legal or illegal, is a step forwards or backwards in our culture, are questions for which there are no easy answers. As with so many other deep human problems, it is likely there will never be one right solution to the question of abortion.

Questions and things to do

Visit your nearest ante-natal clinic. Make notes of all the work done to keep the mother and unborn baby healthy.

Ask your own mother about her feelings during pregnancy and birth.

Do you think a father should be present at the birth of his child? Make a list of all the reasons for and against his being with his wife.

Find out all you can about the work of a midwife.

1. What is fertilization? Explain clearly where it happens.
2. What is meant by 'implantation'?
3. Explain as clearly as you can why the first months of life in the uterus are so important in the baby's development.
4. In what ways can a mother make sure she is fit and healthy before and during her pregnancy?
5. Using diagrams, explain the functions of the placenta and the umbilical cord.
6. If the average period of gestation is 280 days, or 40 weeks, work out the dates when you and four other people in your class were conceived.
7. Describe the three stages of labour. Why do you think the first stage takes so long?
8. What are the most important reasons for a mother to breast-feed her baby?
9. Using diagrams, show the slightly different contents of the uterus of a mother having identical twins to a mother having fraternal twins.
10. Explain why identical twins are always the same sex.
11. Write to the Family Planning Centre for information and leaflets on contraception.
12. Sketch on the outline of the pregnant woman, the position of the foetus in her womb.

FAMILY PLANNING

Method	How Safe	Who uses it	How it is used	More facts	Reasons why it might fail
The 'Pill'	Very, very safe.	Woman	There are clear instructions on the packet when to take the tablets.	Simple and easy to take. Must go to doctor or clinic for check-ups on health.	Forgetting to take tablets. Not following instructions.
The Sheath (condoms, durex, rubbers, french letters)	Quite safe. Very safe if used with a spermicide.	Man	Put on erect penis *before* sexual intercourse.	Easy to buy from slot machines in cinemas, chemists, hairdressers etc. Free from clinic.	Not put on before any contact with vagina. Not carefully removed afterwards.
The IUD (coil or loop)	Very safe.	Woman	Put inside womb by a doctor at clinic and stays there.	Can cause heavier periods for some younger women. Once in place, the woman is free from worry.	It's not yet understood why this method doesn't suit a few women.
The Diaphragm (dutch cap)	Quite safe. Very safe if used with a spermicide.	Woman	Put by woman at top of vagina covering entrance to the womb before sexual intercourse. Removed 8 hours later.	Woman is fitted with cap and taught how to use it at the clinic. Also given spermicide.	Not put in correct place at top of vagina. Must always be used with a spermicide.

FAMILY PLANNING

Method	How Safe	Who uses it	How it is used	More facts	Reasons why it might fail
Withdrawal	Very, very unsafe.	Man	The man removes his penis from the vagina before his sperms are passed into the vagina.	Small amount of semen leaves penis at beginning of sexual intercourse. Millions of sperms in it.	High risk of pregnancy as sperms on outside of vagina can swim up to the egg tubes.
Spermicides (sperm killers)	Not safe but better than nothing.	Woman / Man	Creams, jellies, foams or tablets put high up in vagina. Creams put inside sheath before use.	Not safe on their own but very safe when used with sheath or diaphragm.	The vagina is very folded and it is not possible for spermicides to get into each tiny fold.
The 'Safe Period' (Chart method)	Not safe.	Both	No sexual intercourse after egg leaves ovary and is in oviduct.	Clinic will work out a time chart when it is likely there will be an egg in the oviduct.	Very difficult to work out exact safe or unsafe dates. Sperms and eggs can stay fertile for different lengths of time.
Sterilization	Very, very safe.	Man / Woman	Vasectomy. Sperm tubes are cut and tied back. Egg tubes are cut and tied back.	Simple operation for a man. Woman has to stay in hospital for a few days. Only for couples who already have a family.	

Chapter Fifteen

Passing on family likeness

People say:
  You look just like

—your father
      or mother
      or grandad
      or aunt
      or great-grandmother
      or second cousin
      or brother
      or sister.

Fig 15.1

Fig 15.2

Where do our characteristics come from?

You may look at your family and think you don't 'take after' any of them at all! But you do, though in some families it is easier to notice likenesses than in other families. The things we inherit come directly from our parents and they are called **characteristics**. We inherit *racial* characteristics such as the colour of our skin, hair and eyes; the shape of our head, eyes, nose and mouth; the outlines and shape of our face and body. We inherit *family* characteristics such as body build, blood group, intelligence, length of life. Racial and family characteristics can't really be separated but it is important to remember we inherit both.

We know the egg and sperm meet and fuse together at fertilization. And we know the developing embryo is made from that one egg and that one sperm. So there must be something in those two cells which is able to pass on characteristics. That 'something' is the **chromosomes** you read about in Chapter One.

WHAT ARE CHROMOSOMES?
1. They are tiny thread-like chains which carry the instructions about what we inherit.
2. They are in the nucleus of every living cell.

3. There are 46 single chromosomes, or 23 pairs, in each cell except the sex cells.
4. They are made up of a complicated chemical called DNA and proteins.
5. On each chromosome are hundreds of **genes**.

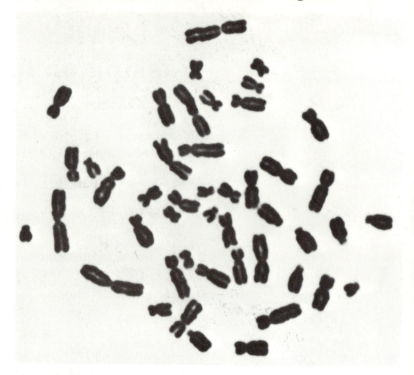

Fig 15.3 Female chromosomes

WHAT ARE GENES?
1. Genes are units of material which make up the chromosome chain.
2. They carry the actual 'words' of the instructions about what we inherit.
3. They are directly responsible for our characteristics.
4. It is believed there are hundreds of genes on each chromosome.
5. The study of genes is called genetics.
We get our characteristics from the genes on our chromosomes.

HOW DO WE GET OUR 23 PAIRS OF CHROMOSOMES?
We get our 23 pairs of chromosomes at the time we are conceived. The sex cells, sperm and egg, are the only cells in the body which have 23 *single* chromosomes each. All other cells have 46 single or 23 *pairs* of chromosomes. Each sperm has only 23 single chromosomes and each egg has only 23 single chromosomes. So when the sperm and egg fuse together at fertilization, we get 23 chromosomes from our father and 23 chromosomes from our mother. The 23 chromosomes from each parent come together and make up our 46 chromosomes, or 23 pairs.

Cell division

Once the 46 chromosomes have paired together in the fertilized egg, the instructions, or 'blue-print', for what we have inherited are set for life. You will remember that the fertilized egg divides into two, four, eight, 16 and so on, till the embryo is formed. We grow and develop by the cells dividing over and over again. The correct name for cell division is **mitosis**, and Figure 15.4 shows what happens during mitosis.

Before the cell divides, the nucleus gets larger.

The membrane around the nucleus dissolves.

The 46 chromosomes double themselves by making exact copies of themselves.

The chromosomes pull apart, 46 moving to the top and 46 moving to the bottom of the cell.

The 46 chromosomes in each half form together into 23 pairs again.

The cytoplasm pulls apart to surround the nucleus of each new cell.

The two 'daughter' cells are exactly the same as the one 'parent' cell before it divided.

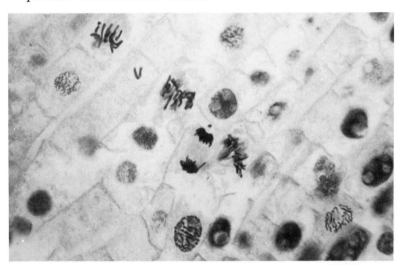

Fig 15.4 Onion cells at different stages of mitosis

Mitosis is easy to understand but you must remember the chromosomes double themselves *before* they divide. This makes sure that each cell divides into two exactly similar cells. The pattern and number of the chromosomes are always exactly repeated. As the embryo develops into the foetus, then the baby, the child, the adolescent and the adult, that very first pattern of chromosomes is still being equally and exactly passed on in the nucleus of the cells. A cell in the tip of your finger, the top of your head, the back of your knee, anywhere in your body, has the exact number and pattern of chromosomes you received when you were conceived.

The only cells which do not divide in this way are the sex cells. The sex cells form in a special way. By the time they are fully formed, mature, they only have 23 single chromosomes each. (If this didn't happen, we would get 92 chromosomes from our parents! It would make a nonsense of the carefully worked out pattern of inheritance.)

We know we inherit all our characteristics. We get exactly half our chromosomes with the genes on them from our father and the other half from our mother. But if you look at it the other way round, you have *not* inherited 50 per cent of your father's or your mother's characteristics. So you cannot be exactly like either of your parents.

Genes

Genes are the 'words' of the instructions about what we inherit. They are responsible for, or **determine**, our characteristics. It is believed each gene has a special place on the chromosome. Most genes are in pairs, one on each of a matching pair of chromosomes, so that most of our characteristics are controlled by pairs of genes.

If we want to study one pair of genes, we show them like this:

The circle is the nucleus.

The ∫∫ are the pair of chromosomes.

The ×× on the chromosomes are the genes.

Gregor Mendel, an Austrian monk (1822–84), was one of the first people to study genes. He called them factors. He experimented with plants to find out what would happen if he crossed the sex cells of two different types. He chose pure-bred tall pea plants and pure-bred short pea plants. The circles in Figure 15.6 show the genes on the chromosomes for tallness H and shortness h. He fertilized the short plants with pollen from the tall plants. He fertilized the tall plants with pollen from the short plants. All the offspring, the first generation, were tall plants. Why weren't the plants all medium-sized? Or why weren't some plants tall and some plants short? And what had happened to the genes for shortness?

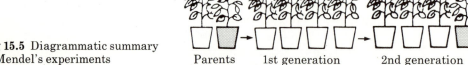

Parents   1st generation   2nd generation

Fig 15.5 Diagrammatic summary of Mendel's experiments

Then he fertilized this first generation of tall plants with one another. Figure 15.5 shows you what happened.

He experimented with large numbers of plants and the results were the same: 75 per cent of them were tall and 25 per cent of them were short. He realized the genes for tallness *and* shortness must be present all the time in the first generation plants, even though all the plants grew tall. Because tallness 'showed through' he called those genes **dominant**. The genes for shortness which didn't 'show through' he called **recessive**. We can study it on these diagrams.

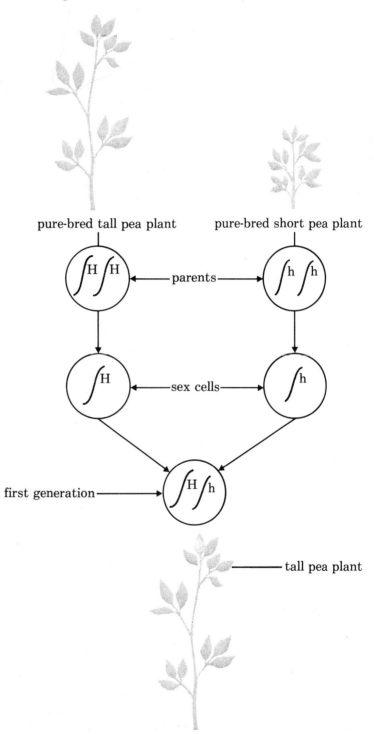

pure-bred tall pea plant

pure-bred short pea plant

∫H ∫H ←——parents——→ ∫h ∫h

∫H ←——sex cells——→ ∫h

first generation ——→ ∫H ∫h

————— tall pea plant

Fig 15.6 First generation

This first generation plant is not pure-bred. Though it is a tall plant, it has two different genes for size on its matching chromosomes. It is now a **hybrid** or mongrel plant. Copy out this diagram and make sure you understand it. In white-skinned people, brown eyes are dominant over blue eyes. Make up your own diagram like Figure 15.6, with one parent having matching genes for brown eyes and the other parent having matching genes for blue eyes. Check your first generation result with those of the rest of the class. Write out in your own words why the child or children will have brown eyes.

In the second generation (Fig. 15.7), we have two plants with matching genes and two plants with non-matching genes, *even though we can see* three tall plants and one short one. What we can *see*, what shows through, the physical type of the plant is called the **phenotype**. What we *can't see*, what the gene type really is, is called the **genotype**. The first tall plant and the short plant have a genotype and phenotype that are the same. They are pure-bred. The second and third plants have a different phenotype from their genotype. They are hybrid.

Make sure you understand this. Copy out the diagran for the second generation until you can draw it without checking back to the book.

Once we know the genotype of a plant, or an animal, or a person, and once we know which genes are dominant or recessive, we can work out the chances of certain genes being passed on to the next generation. And we can understand why certain characteristics can miss a generation and reappear in the second generation.

You already know the gene for brown eyes is dominant over the gene for blue eyes. You have already crossed the genotype for brown eyes with the genotype for blue eyes. Now, without checking back, try to draw a second generation diagram for eye colour from the result of your first crossing. Remember, the offspring had brown eyes but they had different genotypes to their phenotypes. Check your results with those of the rest of the class. Write out in your own words why two brown-eyed parents may have a child with blue eyes. What are the chances of this happening?

Curly hair is dominant over straight hair in white-skinned people. Red and blond-coloured hair is recessive to dark-coloured hair. Practise your diagrams by working out first and second generation charts for these characteristics, starting with parents having the same genotype as phenotype.

However, most of the characteristics we inherit are controlled by *more than one* pair of genes. For example, a great number of genes control the size and shape of the nose. Figure 15.8 is interesting to look at but you cannot

second generation

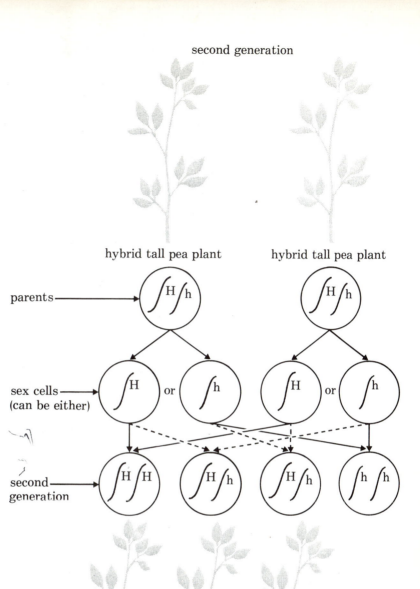

hybrid tall pea plant   hybrid tall pea plant

parents ⟶

sex cells ⟶
(can be either)

H   or   h   H   or   h

second ⟶
generation

HH   Hh   Hh   hh

Fig 15.7 Second generation

work out any charts for inheriting characteristics from it. Where a lot of genes control one thing, it is not possible to find the genotypes needed to begin a diagram.

The famous family tree shown in Figure 15.9 will remind you we inherit mental characteristics as well as physical ones.

Not all genes are dominant or recessive. And not all characteristics we inherit can be so clearly shown as in Mendel's diagrams. Not enough is yet known about skin

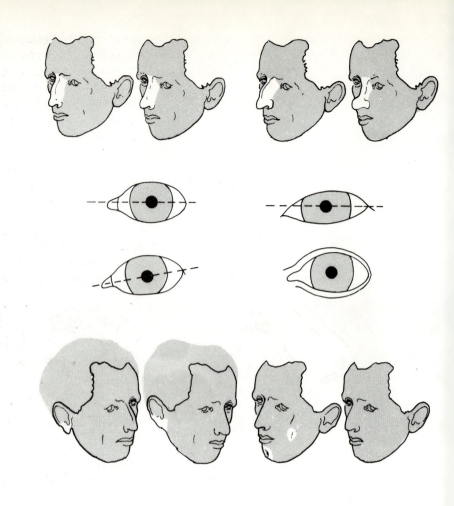

Fig 15.8 Many genes control our looks

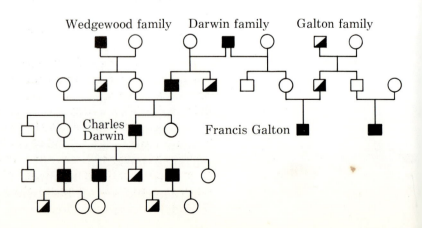

Fig 15.9 A famous family tree. Find out about the work of Charles Darwin

colour, but we do know the dream of mixing the races so we end up with 'a coffee-coloured world' is, in fact, just a dream. It doesn't work like this in practice. The genes which control the melanin, the dark pigment in our skin, are always there in the genotype and can reappear in the second or later generations. It is believed many different genes control skin colour.

The sex of a baby

To sort out which parent's sex cell decides the sex of a baby, we must go back to the chromosomes and look at them more closely.

Notice that all the chromosomes in Figure 15.10b are in matching pairs except the last pair in the man. They do not match. They are the sex chromosomes and are called the X and Y chromosomes. The woman's sex chromosomes do match. They are called the XX chromosomes.

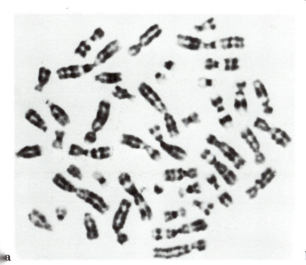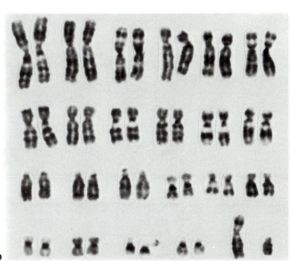

a   b

Fig 15.10 a Male chromosomes
b Male chromosomes sorted into pairs

Remember that the sex cells have only half the number of chromosomes that ordinary cells have. This means that half the man's sperm will have an X chromosome and half will have a Y chromosome. Since the woman's sex chromosomes are both the same, all her eggs will have an X chromosome. If a sperm with an X chromosome fertilizes an egg, then the embryo produced will have two X chromosomes and will be a girl.

If a sperm with a Y chromosome fertilizes an egg, then the embryo produced will have one X and one Y chromosome and will be a boy. A Y chromosome will always produce a boy.

The sex of a baby is set at conception, at fertilization. The man 'decides' the sex of the baby – though, of course, he does not know which sex chromosome will be on the sperm that fertilizes the woman's egg. Each time an egg is fertilized, there is an equal chance of the baby being a boy or girl.

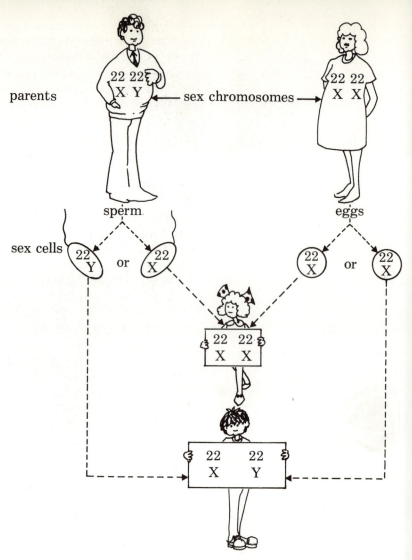

parents   ← sex chromosomes →

sex cells

sperm

eggs

Fig 15.11 The chances of a baby being a boy or a girl are 50:50

In most countries, more baby boys are born than girls. The numbers are about 105 boys to 100 girls for white people and 102 boys to 100 girls for negro people. Why this slight difference happens isn't known. But it is true there is a slightly higher risk of baby boys and men dying than there is for baby girls and women.

Sex-linked characteristics

There are certain recessive genes which show through in men but not in women. How does this happen? Why, for example, are some men red-green colour blind and women are not, especially when the gene for red-green colour blindness is recessive?

The gene for red-green colour blindness is on the X chromosome. In a woman the two sex chromosomes are the same size. She may also have the gene for red-green colour blindness on one X chromosome, but she also has the gene for normal sight on the other X chromosome. The gene for normal sight will dominate the gene for

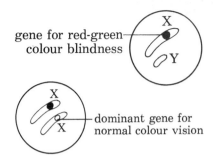

gene for red-green colour blindness

X
Y

X
X
dominant gene for normal colour vision

Fig 15.12

colour blindness. But she can still pass this gene on to her children, and so she is called a 'carrier'.

Now look at the different sizes of the XY chromosomes in a man. You can see that, because the Y chromosome is smaller, there is no place for a matching gene for normal sight on it. When a boy inherits the recessive colour blindness gene from his mother, he has no normal matching gene to dominate it. He will be colour blind.

The same thing happens for any genes on the parts of the X chromosome which have no matching parts on the Y chromosome. Even though they are recessive, they will show through. Can you work out why characteristics such as colour blindness are called **sex-linked characteristics**. Haemophilia (page 37) is another example of a sex-linked gene which is carried by a woman and causes the illness in her sons. Haemophilia is also a disease which comes from a mutation of the genes.

Mutations

The word mutation comes from the Latin 'mutare', 'to change'. A mutation is a sudden change in a gene or a chromosome. It doesn't happen very often. The gene or chromosome is 'shaken up'; it is altered or changed in some mysterious way. Some mutations are caused by outside things such as heavy doses of X-rays, certain chemical compounds and electromagnetic radiation. Some mutations just happen and we don't understand the reason why. You can understand that when a gene changes, mutates, the characteristic it controls is also likely to be changed. If a dominant gene mutates, the characteristic is bound to be changed. Once a mutation happens, the new pattern of genes is passed on in the sperm or egg cells.

A gene or chromosome mutation can be helpful or harmful. In man, they are usually harmful. Can you imagine how difficult life is for boys who inherit the mutated gene for haemophilia? From birth onwards, they must be protected from bumps and bruises which can start internal bleeding. And even a small cut on the skin will cause bleeding which is very difficult to stop. A person suffering from haemophilia can now be treated with a special chemical which helps the blood to clot. But there is *no cure* for this disease, nor for *any disease from a mutated gene*, because each cell in the body will have the mutated gene.

If we look at the chromosomes of a mongol and match them up, like Figure 10b, we notice an extra chromosome on one of the 'pairs'. We don't know why an extra chromosome suddenly appears like this. But we do know that the older a woman is, the more likely she will be to have a mongol baby. Mothers over 40 have a much higher risk than younger mothers.

This extra chromosome throws the whole delicate work

Fig 15.13 A mongol child

of growth and development out of balance. It upsets and interferes with the proper development of the mind and the body. Physically, the mongol child may be stunted, may have a small round head, upward slanting eyes, very short nose, an underlip which juts out and a groove down the tongue. Mentally, the mongol child may be able to learn simple things like washing, dressing and feeding itself. But some children are so damaged by the extra chromosome that they are not able to learn and have to be looked after all the time. However, mongol children are usually happy, cheerful and lively and they are very loving towards their family.

They were called 'mongols' because they have slanting eyes with folds of skin on the inside eyelids like the Mongolian people. This was the name given to them from the eye shape. They have no more links with the

Mongolian people than with any other group of people. The correct name for this disease is **Down's Syndrome**, though most of us still use the word 'mongol'. Mongols are born in equal numbers all over the world. For every 1000 births about six babies will be mongols. It is very rare that a mongol woman has a baby, but if she does, there is a 1 in 2 chance her baby will also be a mongol.

POINTS TO REMEMBER
1. Mutations happen *at random*, by chance.
2. They do not happen very often.
3. We know a few of the 'outside' causes.
4. In man, they are nearly always harmful.
5. The diseases cannot be cured, though many can be treated.
6. The new pattern of genes, or chromosomes, is passed to the children.

Inherited diseases

The diseases which are passed in the sex cells of parents to their children are called **hereditary**. People with a hereditary disease in their family are naturally anxious not to pass it on to their children. They go to a Genetic Counselling Clinic to find out what the chances are of the children being affected. Many people find the risk is so slight they can start their family without worry. Others find the disease in their family is *not* an inherited one. But a few people are told there is a high risk of passing on the disease. A dominant gene carries the risk of 1 in 2 children being affected. Two recessive genes carry the risk of 1 in 4 children being affected.

What should this third group of people do? Especially if their disease is a terrible physical or mental handicap? Would it be sensible or fair for them to have a handicapped baby? Should they be sterilized (page 198) so there is no chance of them passing on the disease? If one or both future parents are badly handicapped, will they be able to give the handicapped baby the extra loving and care it will need? Do they have the right to take the risk and have children? Discuss this now in class. If you have any secret fears that you may have a harmful gene on your genotype, go to a Genetic Counselling Clinic. They will research into your family history and tell you whether you are carrying an inherited disease of not. If you are, they will work out the chances of your children being affected. Don't be afraid to go to the clinic. A lot of diseases we think are inherited aren't inherited at all. But if you find you are carrying a harmful gene, it is much better you know now. You can make plans for your future to suit you. Not many people want a handicapped baby. You will need to be a very special parent to give a handicapped baby all the love and attention it will need.

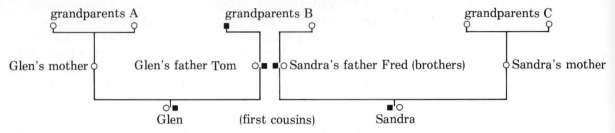

grandparents A   grandparents B   grandparents C

Glen's mother   Glen's father Tom   Sandra's father Fred (brothers)   Sandra's mother

Glen   (first cousins)   Sandra

Fig 15.14

Glen and Sandra are first cousins. Grandmother B has a recessive gene for a terrible disease on her genotype. No one knows about it. Glen and Sandra have fallen in love. They want to marry. They plan to have three children. Their parents are against the marriage. They know it isn't wise to marry in your own family. They have asked you to explain to Glen and Sandra the risks they might be taking.

What advice would you give Glen and Sandra? Write out clearly and simply what you would say to them and, using diagrams, explain the chances of passing on a terrible disease.

Things we don't inherit

DISEASES

A lot of the **congenital** (present at birth) diseases are not inherited at all. They are caused by germs or drugs passing to the baby in the womb during the 9 months of pregnancy. They are caused by the mother not eating the right foods so the unborn baby has a deficiency disease. A lot of congenital diseases are caused by the environment, which is the mother's body.

Syphilis is an example of an infectious disease which will damage the unborn baby. The germs of syphilis pass through the placenta and attack the developing foetus. So, if the mother has syphilis when she gets pregnant or if the father passes it to her while she is pregnant, the unborn baby will be damaged by it. (As syphilis is such a horrible disease, all pregnant mothers now have their blood tested for it.) People still believe that syphilis is an inherited disease. It is not. It is carried by germs in the mother's bloodstream. The germs cross the placenta and the unborn baby 'catches' it from the mother. This is true of all the infectious diseases which can get through the placenta. *We don't inherit infectious diseases, we catch them from our environment.*

ACQUIRED SKILLS

The things we do to improve ourselves are not passed on to our children. The skills we **acquire**, what we make of ourselves mentally and physically, cannot be passed on to our children. All this father's bodybuilding and the

Fig 15.15 This man has developed large muscles: his wife is a professor of history

mother's long hours of study do not in any way change the genes the children will inherit.

It seems a waste that we can't pass on our skills. But then, we can't pass on our lack of skills either. If this father became an alcoholic or the mother was too lazy to train her mind, it still would not affect the genes the children will inherit. We cannot inherit 'good' or 'bad' skills from our parents.

Why is it then, that children from steady, hard-working parents usually do better than children from unreliable, lazy parents? Can you make a guess at the answer now? Think carefully about it.

Heredity and environment

Identical twins come from the same egg and the same sperm. So they must have exactly the same pattern of genes on their chromosomes. They must inherit exactly the same characteristics. We expect identical twins to develop in the same way and to be very much like each other, physically and mentally, even when they are grown-up. But if they are separated when they are babies and brought up in different families, we can expect them to be different from each other in lots of ways.

What will happen to a boy if he has the genes for tallness but is not given a proper diet when he is small? What will happen to a girl if she has the genes for cleverness but is not taught to read and write? The environment which we grow up in has a strong influence over the things we inherit. A 'good' environment helps us to develop what we have inherited. A 'bad' environment stops us developing things we have inherited.

WHICH IS MORE IMPORTANT, OUR HEREDITY OR OUR ENVIRONMENT?

There is no answer to this question as both heredity and environment go to make up the sort of person we become. We cannot say that one is more important than the other. They both influence each other and we are a mixture of this influence.

There is not much we can do to control the genes our children will inherit from us. We can be very careful choosing our marriage partner and we can visit the Genetic Counselling Service if there are any harmful genes in our family. But we can't really do anything more. However, we can do a great deal about the environment we give to our children. We can make sure it is a 'good' one, so our children have a chance to develop the abilities they may have inherited.

Questions and things to do

1. What are chromosomes?
2. What are genes?
3. How many chromosomes do we inherit from our mother?
4. Learn exactly what happens during cell division, mitosis.
5. Explain as simply as you can what is meant by 'genotype' and 'phenotype'.
6. Make sure you can draw the diagrams of inheritance of one pair of genes with (a) pure-bred genotypes and (b) hybrid genotypes.
7. Right-handedness is dominant over left-handedness. Explain why parents who write with their right hand may have a left-handed child.
8. Using diagrams, explain how the sex of a baby is determined.
9. Sex-linked characteristics 'show through' in sons but not in daughters. Can you explain why this happens?
10. Learn 'Points to remember' about mutation.
11. What is the work of the Genetic Counselling Clinics?
12. Which do you think is more important, heredity or environment?

Chapter Sixteen

The first 5 years of life

A new-born baby weighs about 3.5 kilograms, is 50 centimetres long, and has a large head for his body size. He may be wrinkled and thin, smooth and plump, or somewhere in between. He may be calm and peaceful, or restless and lively. Whatever he is like he has arrived, he is helpless, he belongs to his parents and depends on them for all his needs. 'He' may be a boy or a girl and throughout this chapter he, his, him refer to a baby of either sex.

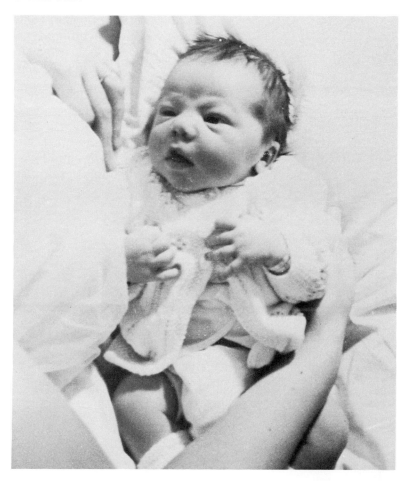

Fig 16.1 This 2 day old boy or girl depends totally on his or her parents

Maturation

A baby's rate of growth and development will depend on the genes he inherits from his parents. New cells will grow, tissues, organs and systems will develop, according to the instructions of his genetic make-up. This will decide when and how the baby first smiles at his mother, or first reaches out to touch a toy. This will decide at what stage he crawls, stands, laughs, runs, talks, plays, makes friends, and so on. A baby's stages of development are inbuilt and they will happen at the *right time for him*. Each baby is an individual. Each baby has his own rate of **maturation**, his own 'time clock' for growth and development.

However, a baby's maturation can be slowed down or **distorted** by a bad environment. During the first 5 years of life, the baby's environment is made up of his mother and father, his family and his home. If this environment is bad, if the people looking after the baby are cruel or careless or neglectful or unloving, the baby will suffer in most areas of growth and development. Turn back to pages 150–153 and re-read them.

To understand a baby's needs, we will divide maturation into four kinds: physical, mental, social and emotional. But we must remember they are all linked together. And each stage of development will depend on the success of the last stage.

PHYSICAL NEEDS

At first, a baby's needs are nearly all physical. He needs food, warmth, comfort and sleep. He needs to be held gently but securely. He needs to hear soft soothing sounds, not harsh sudden noises. He needs to be close to his mother, to feel her warmth and love and caring. He needs to feel pleasure through his skin so he can learn to enjoy his body.

Fig 16.2

As he gets older, he will learn to use his muscles. He will need freedom and encouragement to try out new movements. He will need praise when he tries out new skills. A toddler gets many knocks and bumps while he is learning to control his muscles and to balance. He may need cheering up and lots of comfort. Parents have to protect their small children from accidents without making them timid or afraid to do more exciting things.

MENTAL NEEDS

A baby learns through his senses. He has a very sensitive mouth and during his first year he will put anything and everything in it! He seems able to learn about the feel of things through the skin lining his lips and mouth. He likes to look at brightly-coloured moving things and he especially likes to look at his mother's and father's face. When he is in his pram or cot, make sure there is something interesting for him to look at. He likes to hear the voices of his parents and to copy the sounds they make. Talk to your baby as if he can understand every word you say. He needs to hear words, hundreds and hundreds of words, before he is able to talk himself. This

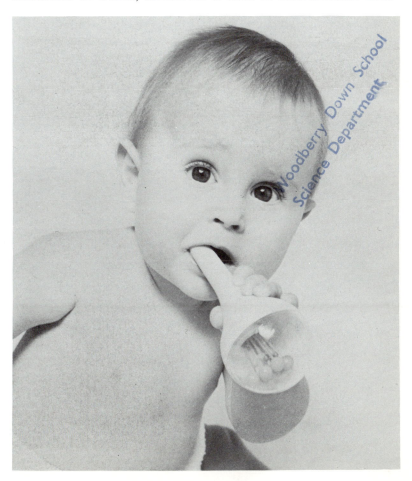

Fig 16.3

will help him learn to speak. All small children need to be talked to, played with, read to and given lots of interesting things to help their minds develop. A child's mind is open and eager to discover his world. He learns more in his first 5 years than at any other time in his life. If he is helped to learn through games and play, then learning will be exciting and full of adventure.

EMOTIONAL NEEDS

We all need to feel loved and secure, to have a close tie or **bond** with one other person, to belong to a family or group of people, to be an important person in the family or group, to be allowed to be individual (to be ourselves) with the person or persons we love. These are our most important and basic emotional needs. A baby and small child has exactly the same needs as the rest of us. At first, he is completely dependent on his mother or whoever is 'mothering' him. He needs to learn **basic trust** in his mother, to be *certain* she will come to him and look after him when he needs her. If she neglects him, if she leaves him alone when he is crying for her, he will learn a basic distrust of her. This makes him feel unloved, insecure, anxious and afraid. It may affect his mental health as he gets older. Satisfying a baby's emotional needs is one of the most important functions of parents.

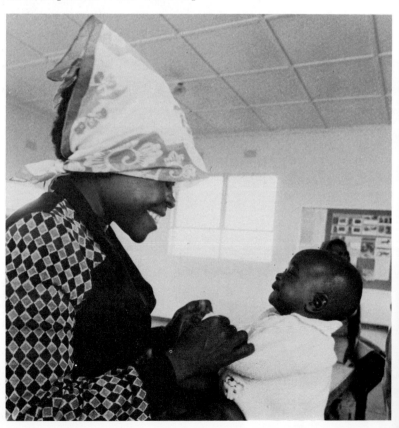

Fig 16.4

218

SOCIAL NEEDS

We are social creatures. We need and want the company of other people. We don't like to be left out, to be on our own for long periods, or to be put with strangers who have no interest in us. A baby needs to be with his mother and family as often as possible. He does not like to be separated from them, to be put in an empty room on his own, or to be looked after by strangers. Most toddlers go through a stage of being very shy and will cry with fear if they are separated from their mother. By the age of 3, some children are ready and need to play with other children of their own age. But other children need only the company of their mother, father and family and are still quite shy when they first go to school. Small children are very different in their social needs. An understanding parent will learn from the child what his needs are.

Fig 16.5

Stages in development

These are average stages in a child's development. But we must remember that with care, love and understanding, a baby's maturation will continue at the right rate for him. A baby is not backward if he doesn't quite fit into an average stage of development. Nor is he advanced if he has made more progress than average. The staff at baby clinics know if a baby is developing at his own proper rate. All babies and small children should be taken to clinics regularly. Not only their health but their stages of development are properly checked.

AT 3 MONTHS

The 3-month-old gurgles, coos and grunts when he is happy. He can be propped up with cushions for a short while so he can see what is going on around him. He tries to grab hold of a toy but his arm and leg movements are still jerky. If a rattle is put into his hand, he can hold it for a time. He may recognize the sound of his parents' voices and can greet them with a smile of joy.

AT 6 MONTHS
The 6-month-old is twice his birth weight and his body is getting larger in proportion to his head. He can't sit without support, but he can hold his head up to see what is going on. He can now reach for a toy and get it. He has a strong urge to play with his hands and learns the size, shape, weight and feel of things by passing them from one hand to the other, by staring at them hard, by putting them in his mouth. He likes rhythm and enjoys being bounced up and down. He will 'chatter' at his parents and squeal with delight when he sees them or hears their voices.

AT 12 MONTHS
The year-old baby may have trebled his birth weight and his body is much longer now. All his muscles are stronger and he has been sitting up for a few months. He crawls or creeps like a bear, on hands and knees, or hands and feet. He may get around by shuffling himself forward on his bottom. He can pull himself up to stand, but gets stuck and can't sit down again. A few 1-year-olds can walk, but this is not usual. He can hold out his arm for his coat and his foot for his shoe. He can pick up a small object between his thumb and forefinger. He waves, plays 'Pat-a-Cake' and may be able to say a couple of words. He is learning to let go, and likes to throw his toys away from him. He enjoys being the centre of attention and shows off to his family. His emotions are developing; he can be jealous

Fig 16.6 3-month-old baby being supported
Fig 16.7 Head raised at 6 months

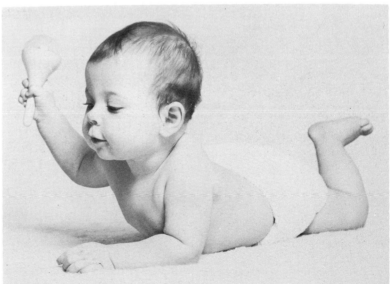

or anxious as well as loving. He knows when his parents are cross or upset. He may try to cheer them up by smiling, laughing or copying them. He is great fun and very interesting company.

AT 18 MONTHS

The 18-month-old has reached the toddler stage. He looks flat-footed and bandy-legged in his new upright position. He needs to practise his new walking skills and spends a lot of his waking time hurrying from one place to another. His movements are not well co-ordinated as he learns to push and pull his toys and the smaller pieces of furniture around his home. He can just manage to throw a ball without falling over. He can point to three parts of his body and can build three bricks on top of one another without knocking them all down. He can only understand his own wants and likes to have his things in the right place. Any change in his routine frightens him and he may lie down, kicking and screaming till he gets what he wants. An 18-month-old is not naughty; his mind has not developed enough to be naughty. When he screams or kicks, he is trying to show his parents how he feels.

THE 2-YEAR-OLD

The 2-year-old has more strength in his back and legs so he looks more graceful when he runs about. He can bend down and pick up a toy without falling flat on his face and he can climb stairs, one step at a time. The 2-year-old can point to four parts of his body and build six bricks. He talks to himself in short sentences, 'Peter take ball' as he plays. He talks to his parents, 'Peter Daddy Walk Out', and loves to be told stories about himself. He likes to play on

Fig 16.8 Crawling at 12 months
Fig 16.9 Almost a toddler

his own as he has not yet learned how to share with other children. He will copy what his parents do and likes to help set the table, make the beds and do other housework. He can now feel shame when his parents are cross with him but this does not help him to do better. He will blame the cat or his toys for whatever has gone wrong. Don't expect 'good' behaviour from a 2-year-old. He needs *help* when he is in difficulties. He cannot understand punishment. He is very loving to his family though still rather shy with strangers.

Fig 16.10 2-year-old with bricks

THE 3-YEAR-OLD

The 3-year-old has much greater control of his movements and far better balance. This means he can be more sure of himself. A sociable child will enjoy going to a playgroup and will no longer be afraid to be left without his mother for a short time. He is proud that he can dress himself, use the toilet and wash his hands without help. He is friendly and wants to please.

He can use his wrists and fingers more skilfully; he can build a tower of nine or 10 bricks and can paint in strokes with his fingers or a brush. He can begin to make shapes from clay, sand or mud. A 3-year-old will do what he is told, *if* he is asked *kindly* and given a reason why. The more words he understands, the more he can begin to reason why he should be kind, helpful and patient, and why he should try to fit his needs in with the needs of his family: both boys and girls especially enjoy playing with their father. A loved and secure 3-year-old is beginning to develop into a social child.

Fig 16.11 Party for a 3-year-old

THE 4-YEAR-OLD

The 4-year-old is curious about everything he sees, hears, smells, touches and tastes. He is full of questions and needs simple, direct answers which don't confuse him. He wants to learn quickly and is impatient with long or difficult explanations. He is boastful and praises himself. He is bossy with his parents and other children and will tell them off if he thinks they are wrong. He loves to dress up and play-act, using his imagination to act out scenes which interest him or upset him. As his imagination develops, he may get sudden fears which parents don't understand. A 4-year-old is very active, trying out new physical skills and crowing with success. At last, he can stand on one leg without losing his balance. He is learning so much, so quickly that he may forget how to do the simple things he learned when he was three. Although the 4-year-old seems quite able to look after himself, he still needs help and encouragement and lots of attention. Rules for his safety are necessary and he will obey them if he is told the reasons for them.

Fig 16.12 This 4-year-old is in danger

THE 5-YEAR-OLD

He is ready for school. He has learned a little about himself and his needs; he has learned a little about his parents, family and friends. Now he is ready to learn about his community, his culture, his world. His personality is made up of what he has inherited and the long five years of physical, mental, social and emotional growth. His personality will go on growing and developing as he matures through childhood. Parents should not feel that once their child goes to school he is 'off their hands', he is 'the school's responsibility'. He is not. School helps a child to learn about the complicated world we live in. But it is the home, the *love* and *understanding care* a child gets from his family, which always has the greatest effect on his growth and development.

Fig 16.13 A confident 5-year-old

Health and hygiene

All through this book, special mention is made of the care of babies and small children. If you are doing a project on child development, go back over the text and copy out the parts to do with the health and hygiene of children. Visit baby clinics, child health centres, nurseries and playgroups to help you with your project. There are many excellent books and leaflets on child development. Write to the Health Education Council or visit your local Health Education Office for more information about the care of children.

Parents are responsible for the health and happiness and proper development of their children. But society is also concerned with the way its future adults are being brought up. Parents are helped in many ways by society.

The midwife is a state-registered nurse with extra training in childbirth. After the birth she will go on caring for the mother and new baby for the next few days.

The health visitor is a state-registered nurse with extra training in child care and general family health. She will visit the family regularly during the baby's first year.

Baby Clinics and Child Health Centres give advice and help to new parents and watch over the proper maturation of the small child.

The family doctor or General Practitioner will make a home visit if the child is too ill to be brought to the **surgery**.

The Paediatrician is a doctor who specializes in children's health and will treat a seriously ill or handicapped child.

The Social Services look after a family in difficulties and give practical help and advice.

The National Society for Prevention of Cruelty to Children protects babies and children from cruel parents. Neglectful parents may have their children taken from them and put into care. Cruel parents may be prosecuted and sent to prison.

Free education and the School Medical Services continue to look after the needs of growing children.
There are many other ways in which society, through the state, looks after the health, happiness and proper development of babies and children.

Chapter Seventeen

Health and disease

Good health is much more than 'not being ill.' In good
health, our mind and body work smoothly together so we
can function at our very best. We have a sense of well-
being and of purpose in life. We are full of energy and we
can work hard. We are of great use to the community.
In good health we are at ease with ourselves, both mentally
and physically.

Disease means we are not at ease with ourselves.
Something is wrong with our body or our mind. They are
not working smoothly together so we cannot function at
our best. We need to be looked after and supported by the
healthy people in the community. In poor health, we are at
'dis-ease' with ourselves, either mentally or physically.

Minor ailments are little things which go wrong from
time to time. They include the aches and pains we all feel
now and then. Minor ailments are usually caused because
we have been careless with our health. For example, we
get a headache when there is too much noise, we get
indigestion if we eat unripe apples, we get a stitch if we
over-exercise and we get stiff if we under-exercise. Minor
ailments are usually warnings or protests against upsetting
the smooth working of the body. They teach us to be more
sensible in the future. Think back to the last time you had
a minor ailment. What caused it? Could you have
prevented it? We have far more control over our health
than we realize.

Fig 17.1

How do we know if we are ill?

This may seem a silly question, but we do sometimes get
muddled ideas about our health.

A few people worry and fuss over their minor ailments.
(We all do this at times when we want a bit of extra love
and sympathy.) Others read all the medical books and
then decide the spot on their arm is leprosy or their
headache is a brain tumour! They are suffering from
hypochondria, imagining they have all sorts of dreadful
diseases. Other people refuse to believe they are ill even
when they are in a lot of pain or have definite signs of
disease. They won't go near a doctor till they are so ill they

227

I'LL BE ALL-RIGHT IT'S ONLY A SCRATCH

Fig 17.2

cannot go on any longer. By this time, it may be too late for the doctor to cure them.

We have to work out a sensible balance between over-fussing about our health and ignoring it completely. *We* cannot decide if we are ill. A doctor is trained to **diagnose** disease. We are not. But we can, and must, decide for ourselves the stage at which we *feel ill* and need a doctor's advice. We can, and must, try to work out the difference between feeling 'under the weather' for a day or so and feeling so unwell that we cannot work or play properly. Remember, it is better to be safe than sorry. It is better to go to the doctor or the Health Centre than to worry about your health. **Preventive medicine**, stopping things going wrong, is the most important part of health care.

Signs and symptoms of disease

The **signs** of an illness are what we can *notice*. We can see the spots of measles, the deformed legs of rickets. We can hear the wheezing breath of the person with asthma or bronchitis. The **symptoms** of an illness are what we *feel*. We feel the hot shivers and aching bones of 'flu. We feel the tiredness, weakness and loss of energy of anaemia. Many diseases have the same signs and symptoms. You cannot diagnose a disease unless you are trained to do so.

A doctor will always ask about your symptoms. Answer clearly, simply and truthfully. Don't expect him to guess. He cannot make the correct diagnosis if you are too shy or too embarrassed to describe how you feel. Babies and small children are not able to tell us their symptoms. This makes it much more difficult to find out what is wrong. Parents can be a great help if they tell the doctor all the differences in the baby since it became unwell.

Types of disease

The most usual diseases are the **infectious diseases** which are caught and then passed from one person to another (page 240). But there are many other types of diseases which are not infectious. And there are dangers, too, in our environment which cause much suffering and pain.

We can't really say that accidents and injuries are diseases. But many of them cause illnesses which last for the person's lifetime. Each year, thousands of people die and many thousands more are crippled for life accidentally. Do a full project on *one* type of accident, e.g. 'Death on the roads', 'Accidents in the home', 'Fire', 'Drowning'. For information on your project write to the Royal Society for Prevention of Accidents (RSPA). Enclose a stamped addressed envelope. Find out the numbers of deaths and the numbers of injured, what are the main causes of the accidents and, most important, the ways we can prevent them happening.

Study First Aid from a properly trained instructor.

Do a full project on occupational diseases.

228

Prevention is better than cure

It is sad but true that many illnesses we suffer need never have happened. They could have been prevented. The boy who is crippled after a motor-bike accident, the man who has a heart attack (page 42) and the child with brain damage from lack of the right food, are all examples of tragic illnesses which need never have happened. How do you think these things could have been prevented?

In the land of Erewhon, any person found guilty of having a minor ailment had to pay a fine. If you had a serious illness, you were sent to prison until you got better or you died! Not looking after your health was a crime

TYPES OF DISEASES		
Name	**Some causes and effects**	**Prevention**
Hereditary disease	A fault in a gene or chromosome can cause mental or physical handicap in the baby	The Genetic Counselling Clinics will advise and help families with an inherited disease.
Congenital disease	An infection or illness in the pregnant woman may damage her unborn baby.	The Ante-Natal Clinics look after the health of the mother and her unborn baby.
Deficiency disease	A diet lacking in the right foods causes scurvy, rickets etc. Some babies suffer brain damage.	Education about correct diet, especially for the young and old. More help given to underdeveloped countries to grow their own food.
Accident and Injury	Lack of care or taking risks with safety causes wounds, burns, crippled limbs etc. Home and street are the most unsafe places.	More care taken to make homes safer, especially for young and old. Laws passed for safety on roads, at work etc.
Mental Illness	Stress, especially in early childhood, may cause mental ill-health in adults. Many other causes.	More education for parents-to-be on the needs of the baby and child. Understanding the importance of a happy home life.
Endocrine disease	Too much or too little of a hormone upsets proper functioning e.g. Not enough growth hormone causes dwarfism.	Regular check-ups at Child Health Centres will result in early notice and early treatment.
Cancers	Most causes not yet known. Changes happen in cell division, cells grow into tumour. *Carcinogens* are chemicals which produce cancer e.g. tars in cigarette smoking.	Avoid known carcinogens. Stop smoking. Cancer *can* be cured especially if found at early stage.
Degenerative (wearing out) disease	As we get old, cells and tissues wear out. Poor sight, hearing, balance, Rheumatism, brittle bones etc.	Old people need to feel wanted and useful. Have important role as grandparents. This helps them to stay active much longer.
Occupational diseases	Work with sand, coal, glass, asbestos, chemicals causes lung disease. Businessmen suffer heart attacks from stress.	Protective clothing and masks must be worn at all times. Businessmen reduce stress, take exercise, eat less.
Behaviour disease	Alcoholism, drug addiction, lung cancer and heart attacks from smoking are examples of *self-inflicted* diseases.	More education on dangers to health and risk of easy addiction.

against yourself and your society. *Erewhon*, or Nowhere, was a book by Samuel Butler you might like to read.

Nowadays, the idea of treating a sick person as a criminal is cruel and unbelievable. There are so many diseases which we cannot prevent. Doctors and research workers devote their whole lives to finding the causes of these diseases. Samuel Butler's idea is interesting only for those illnesses which happen through carelessness or neglect of our health. What do you think of his idea?

'Prevention is better than cure' is not just an easy saying. It is a vital truth we must live by to be a happy and healthy society. We have to be responsible for our own health and for the health of others.

Fig 17.3 Make a list of all the dangers in this room

Chapter Eighteen

Infectious diseases 1: What causes them

Micro-organisms

We share our world with living things that are so small we can't see them unless we look at them under a microscope. We call these tiny living things **microbes** or **micro-organisms**. Some of them, as we can see from Figure 18.1, are very useful to us. Others are harmful to us.

YEASTS ARE MICRO ORGANISMS USED IN MAKING BREAD & WINE

MANY KINDS OF BACTERIA ARE USED TO TURN OUR WASTE INTO HARMLESS MATTER

TWO KINDS OF BACTERIA ARE USED IN MAKING CHEESE AND YOGHURT

MANY KINDS OF BACTERIA CHANGE DEAD PLANTS AND ANIMALS INTO THINGS WHICH ENRICH OR FERTILISE THE SOIL.

THE DRUG PENICILLIN IS MADE FROM A MICROSCOPIC FUNGUS

Fig 18.1 Useful micro-organisms

WHERE DO MICRO-ORGANISMS COME FROM?

There is an old saying, 'Adam had 'em'. They live in the earth, the water, the air. They live on plants and animals. Many, many millions of them live on each human being. They cover our skin, burrowing into the folds and crevices. They make their homes in the hair of our head, eyebrows and eyelashes. Millions of them live inside our mouths, nose and throat. Colonies of them can be found in the intestines, particularly the colon and rectum. A few organs, like the lungs and bladder, are free of microbes. And the unborn baby, protected in the membranes of the uterus, is also free of germs. But once birth starts, and the baby is pushed down the vagina, it meets its first micro-organisms. During the early weeks of life, no matter how carefully it is washed and cleaned, it will begin to get its fair share of them. It is almost impossible to make any adult microbe-free.

We can say that micro-organisms are part of our environment, but it is also true that we are the environment of micro-organisms. When we are fit and healthy, they do not harm us. We have **resistance** to them. A few of them help us, such as the bacteria in the colon which set free vitamin K and some vitamins of the

B group. When we are unfit or unhealthy they may get out of control. When we don't wash our skin, hair or teeth, we get infections such as impetigo, a skin disease, or pyorrhoea, a gum disease. A healthy person who follows the rules of hygiene does not usually have to bother with the micro-organisms which live on him or her.

GROW YOUR OWN MICRO-ORGANISMS

You are going to 'seed' or plant your own micro-organisms on the jelly. *Before you begin*, don't wash your hands. Write out on sticky labels what you will do on each plate and be prepared for what you are going to do before you lift the lid. *When you have finished*, replace the lid immediately. Seal the plate with sticky tape. Fix the label on the bottom side.

Plate 1:   Wipe your finger lightly across the jelly.
Plate 2:   Wash your hands then wipe your finger over the jelly again.
Plate 3:   Wash your hands for a second time then touch the jelly again.
Plate 4:   Pull one hair from your head and wipe it across the jelly.
Plate 5:   Scrape a tooth with your nail and put the scraping on the jelly.
Plate 6:   Breathe heavily over the jelly.
Plate 7:   Just lift the lid briefly and then replace it.
Plate 8:   Don't touch this plate as it is the 'control' jelly.

There are many other things which can be seeded; a little soil, the dust on your desk, tap water, and so on. Wash your hands very thoroughly when you have finished. The sealed plates will be put in an incubator at 37°C and left for 2 days or till your next lesson.

Examine the cultures but do not lift the lids. Draw what you see. Write about all the differences you notice in the eight plates.

Scrubbing up before an operation takes at least 5 minutes. Do you think this is necessary? Each person in the operating theatre must wear sterile clothes, masks over mouth and nose and caps over their hair. From your experiments explain why these are worn.

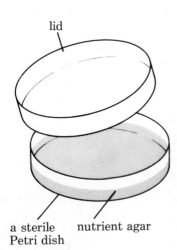

lid

a sterile   nutrient agar
Petri dish

Fig 18.2 A sterile Petri dish

Micro-organisms which cause disease

A micro-organism which causes disease is called a **pathogen**. (We have called them germs so far, but must now study them in detail and use their correct names.) They grow and reproduce on living things, so they are **parasites**. The living things they live on, us and other animals, are called the **hosts**, even though we don't want them and certainly haven't invited them!

VIRUSES

These are the smallest pathogens. They are so tiny they can only be seen by using a powerful electron microscope.

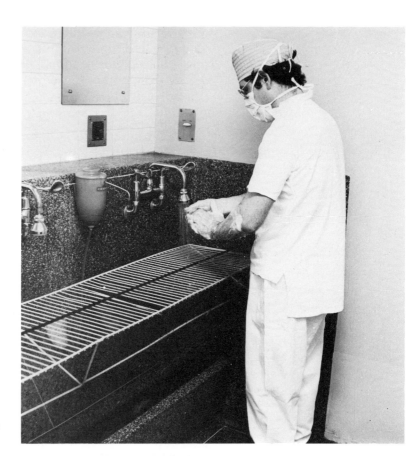

Fig 18.3 A surgeon scrubbing up before an operation

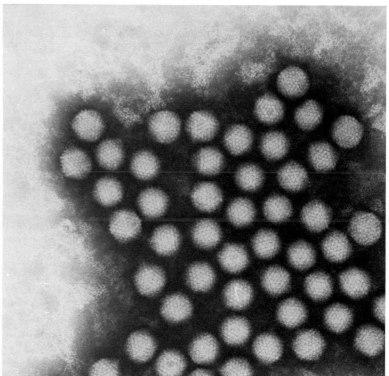

Fig 18.4 Photomicrograph of a group of viruses

Viruses only grow in living tissue, where they multiply rapidly. Scientists can now grow them in hens' eggs so we are learning more about them. Some diseases caused by viruses are mumps, measles, smallpox, yellow fever and influenza or 'flu.

RICKETTSIAE

These are named after Dr Ricketts who discovered them. (They have nothing to do with the food deficiency disease, rickets.) Like viruses, they only grow in living tissue. They are larger than viruses but not as large as bacteria, though they look the same. They live in the intestines of ticks, lice and fleas and cause different types of typhus disease, one of which Dr Ricketts died from.

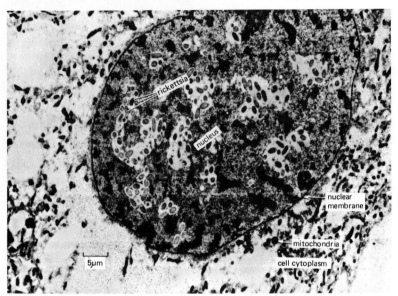

Fig 18.5 Photomicrograph of rickettsia in the nucleus of a cell

BACTERIA

Bacteria are large enough to be seen with a normal microscope. They can be grown in a culture, as you have done, which makes them easier to study. They reproduce by simple division every 20 minutes. Work out how many bacteria you would have in your throat when you woke up if one landed at 6 p.m. the night before. Some diseases caused by bacteria are gonorrhoea, pneumonia, tetanus, cholera and tuberculosis or TB. Some bacteria can make a tough covering for themselves and stay **inactive** (not growing or reproducing) inside these **spores**. The tiny spores can be blown about in the air and settle on dust, clothes or anywhere. They will not be active again till they are blown or moved onto food or an open wound in our skin. Some spores are so tough they are not killed by boiling water, by freezing or by dryness.

FUNGI

Most fungi are useful to man: there are not many which

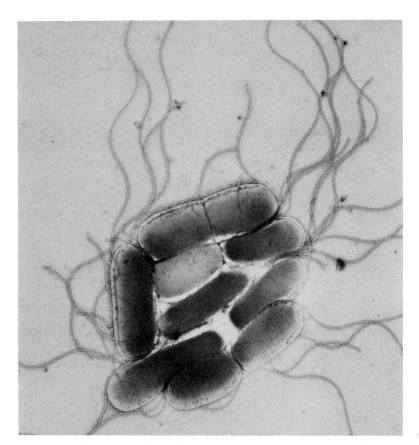

Fig 18.6 Photomicrograph of a group of bacteria

Fig 18.7 Photomicrograph of the fungus that causes thrush

cause disease. We have read about ringworm on page 116. Another kind is **thrush**, a yeast-like fungus which grows in the lining of the mouth. Babies often suffer from it. It can also grow in the vagina and in the rectum.

PROTOZOANS

These are tiny one-celled animals which usually live in fresh water or soil. The amoeba is a good example of a protozoan but it is not pathogenic. But many protozoans are pathogenic and are extremely dangerous to man.

Malaria is caused by a protozoan which lives in the female *Anopheles* mosquito. When we are bitten by this mosquito, the malaria protozoan travels into our blood and liver cells, where it destroys our red blood corpuscles.

African sleeping sickness is caused by a protozoan which lives in the tsetse fly and is passed to us when we are bitten by the fly.

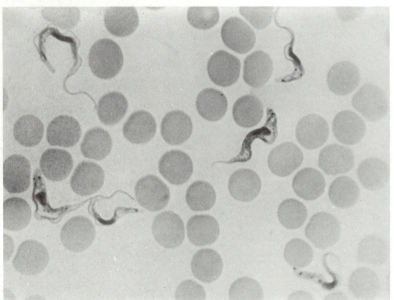

Fig 18.8 Photomicrograph of the protozoa which cause sleeping sickness

Draw one of each type of pathogen and write notes about them. (Turn to page 178 and study the causes of sexually-transmitted diseases. How many different types of micro-organisms cause them? Can you find the same types of micro-organisms as the ones you have just read about?)

Worms

We also share our world with worms. Worms are not micro-organisms. They can easily be seen as they are made up of many millions of cells and are huge in comparison with micro-organisms. Most worms are very useful to man but there are some types which are pathogenic. They spend some or all of their lives inside man so they are also parasitic. It may be difficult to think of worms as causing dreadful disease, but Figure 18.9 will give you some idea of the damage they can do to people in warmer countries.

THREAD WORMS OR PIN WORMS

Thread worms are found in people of all countries. They look like small white threads. The male is 0.4 centimetres long and the female is 1.25 centimetres long. They live

236

Fig 18.9 This man has a disease called river blindness caused by worms

mainly in the large intestine, especially in young children. At night, the female travels down to the anus to lay her eggs. As she wriggles around, she causes great itching and irritation. Naturally, the child scratches itself and the sticky eggs are caught under the fingernails. The child will re-infect itself when it puts food in its mouth or sucks or bites its nails. Thread worms can be seen wriggling about in the waste food, the faeces, or on the paper used to wipe the child's bottom. They do not usually cause disease but they upset and disturb the child at night. They are easily killed by safe drugs which can be bought from the chemist. Every person in the house should be treated as the eggs can dry up and be blown around, landing on dust, food and bedclothes. Nails should be cut very short, hands often washed, especially after using the toilet, and much greater care taken over cleanliness and hygiene in the home.

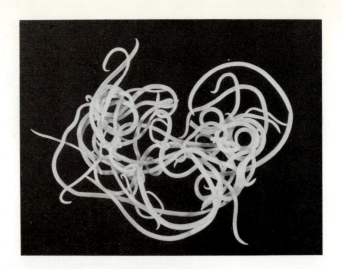

Fig 18.10 Human round worms

ROUND WORMS

Round worms are common all over the world though very
rare in Britain. They look rather like earth-worms, can
grow up to 30 centimetres long and are a shining white or
pink in colour. We, the hosts, catch round worms by
swallowing food which is infested with their eggs. The
tough walls of the eggs are broken down in the intestine
and the tiny worms burrow into the nearest blood vessel.
They are carried in the bloodstream back to the heart and
out to the lungs. For a while they settle in the lungs. Then,
and you may find this rather unpleasant, they pierce the
wall of the lung and, once in the air spaces, wriggle up the
trachea to the back of the throat. The host swallows them
and they return down the digestive tract to the small
intestine. Here they breed and vast numbers of eggs are
passed out with the faeces.

Round worms cause great damage. As they live off the
digested food in the gut, they cause severe malnutrition
and many children die as a result. There are safe drugs
which will kill off the round worm but it is much more
difficult to prevent people from being re-infected. Strict
personal and community hygiene are necessary. Safe
disposal of sewage (faeces and urine) will be studied later
on. All food must be cooked thoroughly to destroy the
eggs. Fresh fruit and vegetables must be washed in clean
water. Food must not be grown in soil near sewage.

HOOK WORMS

Hook worms are common in warm and tropical countries.
They are tiny, about 1 centimetre long, and live in water
for the first part of their life. They get into the human host
by piercing the skin on the feet or ankles of people
walking in infected water or working on damp soil. Their
eggs are also swallowed. They travel, like the round worm,

to the lungs and back to the gut. But they are blood-suckers so they feed off the blood in the intestine wall. They cause serious wounds in the walls of the blood vessels and suck such large amounts of blood that the host suffers from anaemia. The host may be ill enough to need a blood transfusion. Treatment can be given after the host recovers some strength. As the eggs are passed out in the faeces, strict personal and community hygiene are necessary to prevent more infections. Feet and ankles should be protected.

TAPEWORMS

These are the best known of the parasitic worms because of all the horror stories told about them. However, they are very rare in Britain. They are flat, greyish-white and can grow to the full length of the gut! The tapeworm has a small round head with suckers. The suckers hold it firmly in place in the small intestine and prevent it being swept out of the gut by peristalsis. It is made up of segments which grow from its neck. Each segment has its own male and female reproductive organs and can fertilize itself. The end segments in the rectum are full of ripe eggs, any number between 30 000 and 40 000 eggs are in each segment. About six segments are passed out with the faeces each day.

Tapeworms have two different hosts. They live for one part of their life in cattle, pigs or fish, as there are beef, pork and fish tapeworms, and they live the other part of their life in man. Study Figure 18.12 showing the life cycle of the pork tapeworm to help you understand this.

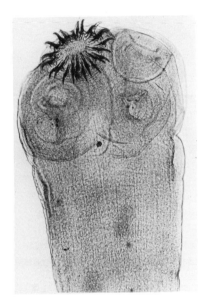

Fig 18.11 The head of a tapeworm

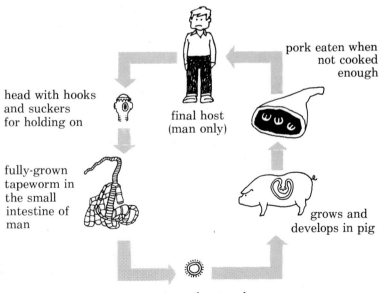

pork eaten when not cooked enough

head with hooks and suckers for holding on

final host (man only)

fully-grown tapeworm in the small intestine of man

grows and develops in pig

egg passed out and eaten by the pig

Fig 18.12 The life cycle of a pork tapeworm

The carcases of newly-killed beef and pork are examined by meat inspectors. If any cysts are found, the carcases are destroyed or put in refrigeration for 3 weeks to kill the cysts. Why should pork always be thoroughly cooked?

There are many other much smaller worms which cause widespread diseases in warm countries. One disease, **bilharziasis**, affects about 200 million people, making them weak, sick and unable to work. The pathogen of bilharziasis lives for part of its life in snails and the other part in man. Like the hookworm, it gets into the person's body from infected water. The eggs are passed out in faeces and urine back into the water.

The disease, river blindness (see Figure 18.9), gets into the person's body by the bite of the black fly which carries the eggs of the worms.

How we get infectious diseases

We get infectious diseases from micro-organisms and worms which are pathogenic to man. They are all parasites because they live on us, the host; they get their nourishment from us and they cause us damage and disease. They are *carried to us* by the air, by water, by food; by plants, animals and insects (re-read pages 114–118 now). They are *passed amongst us* by direct contact, touching infected persons and open wounds; and by indirect contact, unwashed plates, glasses, cups, pots and pans, clothes and bedding; dirty toilets and untreated sewage. Copy out and learn this paragraph. The five main ways in which diseases are spread are shown in Figure 18.13.

by water (eg cholera)   by food (eg food poisoning)   by air (eg common cold)   by insects (eg malaria)   by contact (eg scabies)

Fig 18.13 How we get infectious diseases

In studying the different parts of the body, we have learned how to prevent the spread of infectious diseases by *personal* standards of hygiene. We will now learn more about *community* standards of hygiene.

Chapter Nineteen

Infectious diseases 2: Stopping them spreading

People live together in communities; some in small villages, some in towns and many in cities. The closer they live together the easier it is for pathogens to be spread amongst them. In crowded cities, there are always risks of outbreaks of disease.

To prevent the spread of disease we need safe water, safe food, clean air, and protection against insects and other disease-carrying animals. We need to have a proper system of getting rid of waste: sewage, household and industrial waste.

Safe water

Water comes from rain, snow and hail. As it falls to the ground, it is usually clean, soft and perfectly safe to drink. It collects in lakes and **reservoirs**, where it can be polluted by bacteria and by dead plants and animals. It forms into streams and rivers which will be polluted if they are used as toilets and as dumps for rubbish. It sinks underground into **wells** and **springs**. Surface wells are often polluted by sewage. Deep wells may be free from pollution.

We can make water safe to drink by boiling it for 5 minutes. This will destroy any pathogenic micro-organisms and worm eggs in it. When you are camping, or if you are on holiday and you think the water may be polluted, always boil any water you are going to drink for the full 5 minutes. Never take risks with water for drinking.

Boiled water tastes flat and dead. And anyway, it isn't possible to boil all the water needed for each person. So our supplies of water are made safe by filtration and by sterilization.

FILTRATION
Figure 19.1 is a simple diagram to show you how filtration cleans the water. The micro-organisms, algae, are able to destroy harmful bacteria. The other layers remove any solid bits.

STERILIZATION
After the water is filtered it is crystal clear. To make absolutely certain it is safe to drink, small amounts of

241

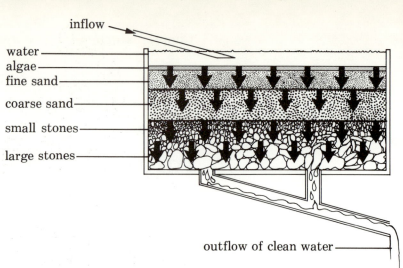

inflow

water
algae
fine sand
coarse sand
small stones
large stones

outflow of clean water

Fig 19.1 A simple diagram to show how water is filtered and cleaned

chlorine gas are added to it. Chlorine is a powerful **bacteriocide**, bacteria killer, and the water is now fit to drink.

The water is pumped to a storage tank and from there is taken by water pipes to the towns and cities. Water, we are told, falls on the just and the unjust alike. It comes as a free gift from the heavens. Why then do we have to pay water rates?

There are still some parts of Britain that are not supplied with piped water. And in country areas all over the world many people get their water from rivers and wells. Imagine you have to carry all your water home in a jar. It will give you some idea of how *precious* water is when you don't have a supply coming out of a tap. And you will also understand how difficult it is to keep clean when all your water must be fetched from the river.

The water in the deep well shown in Figure 19.3 is safe to drink because it has had to drain down through natural

Fig 19.2 An aerial photograph of a water treatment plant

242

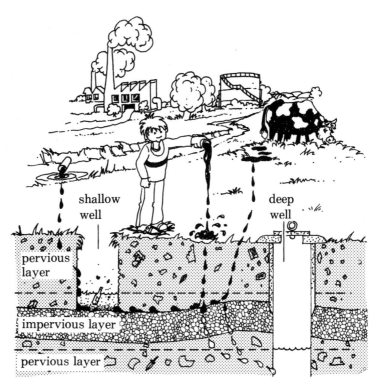

shallow
well

deep
well

pervious
layer

impervious layer

pervious layer

Fig 19.3 The water in the deep well is safe to drink

filters of rock or clay. Notice that no water from the topsoil can get into the well because it is lined with concrete. It is protected from airborne bacteria by the close-fitting lid. If there are cholera bacteria in the surface well, healthy people will get infected and will go on re-infecting themselves.

CHOLERA IS A DANGEROUS HUMAN DISEASE

Cholera bacteria are passed from the faeces of an infected person into water. The person either drinks the infected water or eats food washed in it. Flies also carry the cholera bacteria onto food. The bacteria breed in the intestines and, after a few days, the infected person gets violent diarrhoea (page 94). The patient loses so much water and essential mineral salts that, if he is not treated, he will die. There are now safe drugs for treating cholera.

The methods of prevention are to close down the infected water supply and to bury the infected faeces a long way from drinking water and soil used for growing food. Patients must be **isolated**, separated, as even their clothing will be highly infectious. People who have been in close contact with an infected person must tell the doctor as they may be 'carrying' the disease. Nurses, doctors and people travelling to infected areas may be inoculated against the disease (page 269). India and nearby countries have a very high rate of cholera outbreaks. In Britain, the battle against cholera has been won because of our safe water supply. Any outbreak of cholera is brought from abroad by foreigners, travellers and holiday-makers.

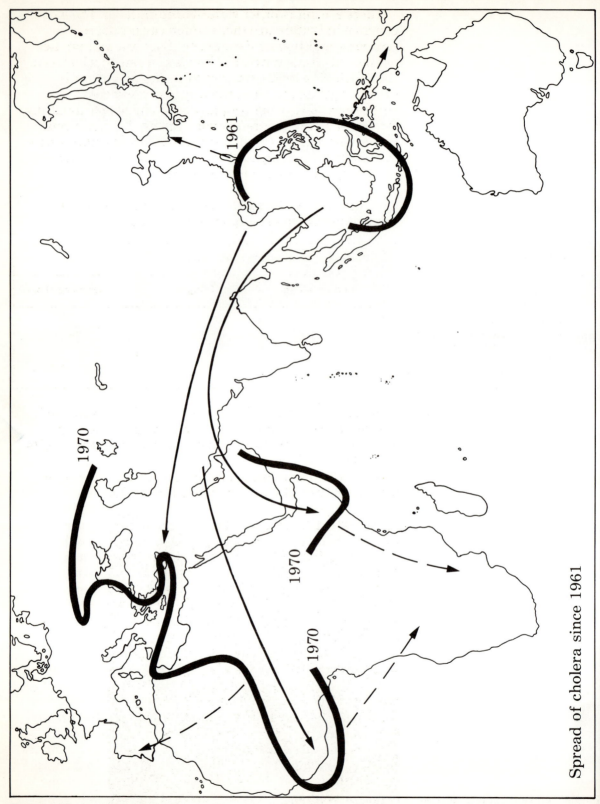

Fig 19.4 The spread of cholera

There are many other water-borne diseases. The most common in Britain are those which cause **gastro-enteritis** and **infant dysentery**. Bacteria and viruses are passed on the unwashed hands of a person who has used the toilet. If he or she touches food, cooking utensils, drinking glasses or cups, the tiny bits of infected faeces will be spread to other people. Cooks and people handling food must be extra careful over hygiene. *Everyone* must wash their hands after going to the toilet. No matter how safe our water supply is, if we have dirty personal habits we will spread water-borne diseases.

Safe food

Do these tests with agar plates to find out the sort of conditions that micro-organisms grow best in.

Some foods	Places to put the plates and results		
	Window-ledge	Fridge	Dark, warm cupboard
a) Scraping of mould from bread b) Scraping of mouldy part from potato c) A tiny piece of boiled meat d) A tiny piece of uncooked meat e) Any dried food you would like to test			

From your results, work out which conditions you think best for micro-organisms to grow in. You can test other foods you are interested in. What do you think the results from plates c and d tell us?

It is rather difficult to get a fly to walk across an agar plate so Figure 19.5 will show you the results.

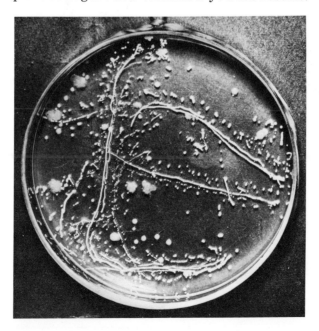

Fig 19.5 An agar plate, incubated after a fly has walked over it

We eat many micro-organisms every day. Most of them are harmless, or get destroyed by cooking or by the acid from the stomach wall. It is quite rare to find food infested with worm eggs in Britain, but we do eat food which has been **spoiled**, made unfit to eat, by the action of bacteria and fungi living on it. Our senses of taste, smell and sight warn us when meat is 'high', milk is sour, jam is mouldy, and so on. But some spoiled foods don't look, smell or taste bad and when we eat them we get **food poisoning**. The other way in which food is made unsafe is when it is washed in polluted water and eaten raw, or when flies, cockroaches and other insects have crawled over it and infected it with their faeces, urine and the micro-organisms on their bodies. And, of course, we ourselves infect food if we have dirty or careless hygiene habits. We need to learn how to **preserve** or keep, store and handle food so we do not make it unsafe for other people or for ourselves.

Fig 19.6 Care must be taken in handling food

THE WAYS IN WHICH FOOD IS PRESERVED

Preservation of food is any method used to keep food pleasant to taste, full of its goodness (nutritious) and free from micro-organisms. The food should be fresh and in good condition before it is preserved.

CANNING

This is a very successful way of preserving food and drink. The food is heated and when hot put into tins which are made of sheet steel, coated with a layer of tin. The tins are sealed to keep out any bacteria. As the food cools, it contracts and this leaves a vacuum at the top of the can. The little hiss you hear when you open a tin is the air rushing into this vacuum.

Fig 19.7 A 'blown' can has bacteria growing in it so the sides or the ends are bulging out. Food from a blown can should *never* be eaten

Look in the food cupboard at home and count how many tins of food are stored. Write down a list of foods you eat which are canned.

BOTTLING

Foods which can be boiled are often bottled at home. The boiling food is put in glass jars and sealed with a glass lid and a rubber ring to prevent bacteria getting in.

FREEZING

Bacteria are inactive in very cold temperatures. Food is frozen very rapidly to −18°C. It can then be transported for thousands of miles in refrigerated ships. Beef from Argentina, lamb from New Zealand and bacon from Denmark are brought to us this way. Frozen foods must be kept frozen till they are ready to be cooked.

Fig 19.8 Dented cans may have been knocked in transport. They are sometimes sold off cheaply because the tin lining may be damaged and bacteria may have got in

Some people have home freezers to store frozen food and to freeze fruit and vegetables from their garden. The family fridge has a compartment for making ice where frozen packets can be kept. Food will only keep for a few days in other parts of the fridge.

We must follow the cooking instructions for frozen foods very carefully. Bacteria are made inactive but *not killed* by freezing, so frozen food must be thoroughly cooked.

DRYING

Bacteria need moisture to be active. If we can remove the water from food it will keep for a long time. Air, vacuum and freeze drying are all ways to remove water from food. Before cooking, liquid is added to these foods and they swell up again. Like frozen foods, they must be thoroughly cooked before eating.

PASTEURIZATION

This process is mainly used to treat milk. The name comes from the scientist Louis Pasteur, who first discovered how to stop wine going sour without boiling it and ruining its taste. He found that if he heated the wine to 50–60°C and

Fig 19.9 Plant for freeze-drying food

Fig 19.10 Pasteurizing milk

kept it at that temperature for 30 minutes, most harmful micro-organisms would be killed and the wine still tasted good to drink. Milk can be treated in the same way, by heating it to 63°C for 30 minutes and then cooling rapidly. Almost all milk is now pasteurized at a higher temperature 71°C for only about 17 seconds before being cooled rapidly. The rapid cooling stops any bacteria which might not have been destroyed from being active, and the milk still keeps its lovely fresh flavour. Pasteurized milk is safe to drink and will 'keep' for a few days if it is not opened and is left in a cool place. Milk which is **sterilized** is boiled at 100°C for 10 minutes, killing all microbes. It will keep for a long time but does not taste quite as nice as fresh pasteurized milk. The standards of hygiene in the treatment of milk are very high. Public Health Inspectors visit farms, bottling plants and dairies to make sure our milk is safe to drink.

FOOD POISONING

Once food is preserved, it is important to store and handle it with care. Food poisoning can be very painful, but except in the cases of babies and small children, elderly or sick people, it does not usually kill.

There are many different kinds of food poisoning. The most common in Britain is caused by the **salmonella** bacteria. When we eat infected food, the bacteria breed in the lining of the stomach and the intestines. In a mild attack of food poisoning we get pains in our stomach and abdomen, and we get diarrhoea. Our body defences help us to get over a mild attack. As the diarrhoea will be full of salmonella bacteria we must take great care over cleanliness after using the toilet or we will re-infect ourselves and others. A severe attack of food poisoning needs medical treatment. Fever, vomiting, diarrhoea and very bad pains all over the abdomen will start within a day of eating the infected food. There is a danger the person will become **dehydrated**, dried out, because water and essential body fluids are lost in the diarrhoea and vomiting. This is especially dangerous for babies, who must be seen by a doctor as soon as possible.

Fig 19.11 Why might these foods go bad?

Figure 19.11 shows the foods most responsible for outbreaks of salmonella food poisoning. Give reasons why these foods might 'go bad'.

There are many other food-borne infections which cause serious disease. Remember, small children should be taught not to eat wild mushrooms or berries. Fresh fruit and vegetables should be washed in case they have been sprayed with insecticides. Cooking pots and pans may cause *chemical* food poisoning if they are chipped or damaged.

Food Health Inspectors are kept very busy making sure our food is safe to eat. Their work includes inspecting the preservation, storing and handling of food in factories, warehouses, shops, restaurants and street stalls.

Clean air

We have already learned about air pollution in Chapter 5 and because it is so dangerous to health, it will be useful if you go back now and re-read the section on respiration.

Some air-borne diseases are tuberculosis, the common cold, whooping cough, sore throat, measles, mumps, diphtheria, 'flu, bronchitis and pneumonia. Air-borne diseases cause infections of the respiratory passages. The bacteria and viruses live in the nose, sinuses, throat, trachea and lungs. It is warm and dark, and there is plenty of moisture and food on the linings of the passages.

When we have a respiratory infection, every time we speak, cough, sneeze or even breathe out, we send thousands of infected droplets spraying out into the air. Anyone within 5 metres is likely to be showered with them! Other droplets may fall to the ground, where the bacteria or viruses dry up, and are then breathed in with moving dust. When we have a respiratory infection we must protect other people from catching it. What things can you do to prevent spreading your cold?

Outbreaks of respiratory diseases are very common in countries with a cool climate. They begin in late autumn and carry on right through winter and into spring. Many elderly people and babies die from them each year.

TUBERCULOSIS IS A WORLD-WIDE DISEASE

Tuberculosis, or TB, is spread by droplet infection and is caused by bacteria. Most people have the TB bacteria but they do not develop the disease unless they are very 'run down' through malnutrition or overwork and worry or they are exposed to more of the bacteria from living in overcrowded infected areas. The TB bacteria attack the alveoli and, if they are not destroyed by the body defences, form **tubercles**. Bacteria multiply inside the tubercle and spread into the bloodstream of the host. Some symptoms of TB are a dry cough, fever, sweating at night, loss of appetite and weight. The patient becomes exhausted and may begin to cough up blood. One hundred years ago, TB was called **consumption**, as the person seemed to burn away or be consumed by the disease. The TB bacteria not only destroy lung tissue. They can invade other parts of the body as well.

TB used to be one of the greatest killers, but there are now drugs which can cure it. Patients are isolated and given plenty of rest and a nourishing diet. The methods of preventing it spreading are by pulling down slums, testing cow's milk for the TB bacteria, and having X-ray units visit schools, offices and factories to **screen**, or check, for early signs of the disease. It is illegal for any person suffering from TB to work in the food trade. Teachers of young children and nursery workers must have a chest X-ray at regular intervals. The BCG inoculation will help to build up an immunity to the TB bacteria (page 269).

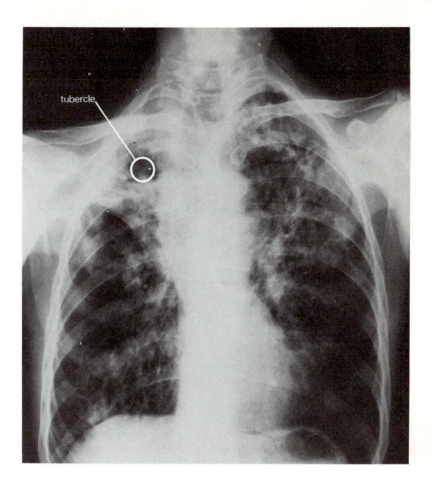

Fig 19.12 X-ray of lungs with a tubercle of TB

Insects and other disease-carrying animals

Animals which carry disease are called **vectors**. Flies, cockroaches, ants, ticks, mosquitoes, tsetse flies, lice, fleas, rats and mice are all animal vectors.

MALARIA

The word malaria means 'bad air' and the disease was given this name because people living near marshy or swampy land thought it came from the bad air over the water. In Britain it was called 'fen ague'; the word **ague** describes the violent shaking which is part of the disease. Great efforts by the World Health Organization have been made to control this disease and you can see the results of their work from Figure 19.13. But millions of people still suffer from it.

Malaria is caused by a protozoan, a one-celled micro-organism, which is passed to humans by the female *Anopheles* mosquito. She pours saliva down one of her feeding tubes into the host so that the blood she sucks up will not clot. The protozoans in the saliva are injected into the host. Thousands of them are taken to the liver where they multiply to millions. They then attack the red blood

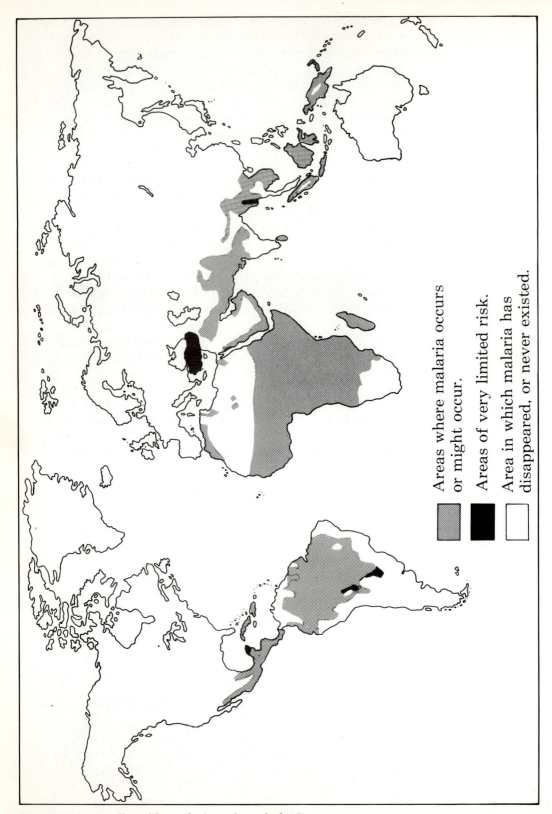

Areas where malaria occurs
or might occur.

Areas of very limited risk.

Area in which malaria has
disappeared, or never existed.

Fig 19.13 Areas affected by malaria at the end of 1975

cells in the bloodstream and destroy them. About 10 days after infection, the patient feels very very cold and gets an attack of the shivers. Suddenly, he feels very very hot, his temperature rises to 40°C, his breathing is rapid and shallow, his pulse rate increases and he has a lot of pain, especially over the spleen. After a time, he begins to sweat heavily, his body temperature drops to normal and he feels better though very weak. In a little while he gets another attack and then another until, without treatment, he dies. There are many different drugs used in the treatment of malaria and there are anti-malaria tablets which can be taken by people going to infested areas.

Ways to control the malaria mosquito

Mosquitoes lay their eggs on still water such as ponds, lakes, swamps, gutters, drains and even jars of water. The larvae hatch out and hang from the surface of the water, getting their air through a small breathing tube. If oil is spread in a film over the water, the breathing tube is blocked and the larvae suffocate. If insecticide is also sprayed, the mosquitoes coming to lay their eggs are killed. If all water in jars is covered with a lid and heavily infested swamps are drained, there will be less places for the mosquito to lay her eggs.

The adult mosquito lives in dark corners, walls and cupboards of people's houses. She comes out at night to get more blood from her victims. A patient suffering from malaria must be protected from other mosquitoes. This stops a malaria-free mosquito becoming infected and carrying the infection to other people.

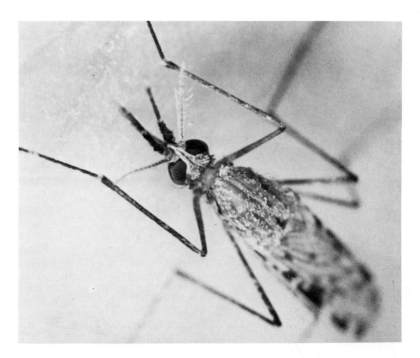

Fig 19.14 A malaria mosquito feeding on a human arm

THE HOUSEFLY

The housefly is responsible for causing the outbreak of many serious diseases. It is common all over the world; wherever man lives. Flies live on faeces, sweet and rotting waste foods, and man provides plenty of these things. Examine a housefly under a hand lens. Draw and label everything you see. Wash your hands really well afterwards.

Body and legs covered with hairs which carry disease.

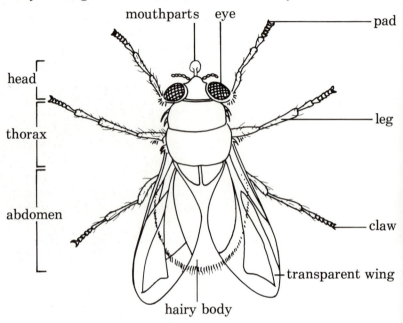

Fig 19.15 Diagram of a housefly

The housefly doesn't live on us as lice and fleas do; nor does it bite us as mosquitoes do. So we think of it as a nuisance but a fairly harmless nuisance. In fact, houseflies can and do carry, *in* and *on* their bodies, any micro-organisms they come into contact with. As they live, breed and feed on faeces and rotting matter, they are perfect carriers of all types of infectious diseases. They have such disgusting feeding habits that any food they land on is likely to be infected.

Feeding habits

Flies can only suck up liquid food through their mouth parts. When they land on some food they spray a part of it with their saliva to make it liquid. Micro-organisms in the saliva infect the rest of the food. They suck up the bit of liquid food and then **regurgitate** it or vomit it back, before sucking it up once more. As they crawl over the food, softening it with saliva, they drop their own faeces on it. Their bodies and legs are covered with hairs which hold dust, micro-organisms and tiny bits of human and animal faeces. These drop off onto the food as the flies clean their bodies. Flies **contaminate** or infect food by:

spraying it with their saliva;
vomiting half-digested food back;
dropping their own faeces on it;
carrying many types of disease on their bodies, legs and
feet.

The female is attracted to rotting scraps of food, dustbins,
rubbish dumps and sewage by the smell. With the tip of her
abdomen she pushes her eggs, about 100 a time, into the
steamy filth. The larvae or maggots hatch out in 24 hours
and feed on the decaying matter around them. Examine a
maggot under a hand lens and draw what you see.

When the larvae have finished growing they form into
pupae. Inside the pupae they change into adult flies. An
adult fly will live anywhere near food.

Fig 19.16

Some other animal vectors are family pets. Dogs, cats,
budgies and hamsters can all pass disease to us. The killer
disease **rabies**, which attacks the nervous system and for
which there is no cure, is carried by infected dogs and
small animals. No pets are allowed into Britain till they
have spent 6 months in special kennels, **quarantine**, and
then been passed as free from rabies. The port health
authorities keep a strict watch to prevent people
smuggling in their pets.

All pet dogs sniff around the urine and faeces of other
dogs and so have dirty muzzles. The child in Figure 19.17
may be getting a serious disease from his beloved pet.

Recently it was discovered that dogs and cats carry a
worm in their intestines which is harmful to us. It is a
round worm, about 10 to 15 centimetres long, and lives in
us in the same way as the other round worms (page 238).
The disease they cause is **toxicariasis**, and it can cause
liver damage, asthma and bronchitis. The eggs are passed
in faeces. Dogs and cats foul anywhere, which is safe in
the countryside but *very dangerous* in towns and cities.
Parks, gardens, playgrounds and streets are fouled with
urine and faeces which attract flies. Small children,
playing outdoors, are easily infected. Adults carry the

Fig 19.17 The family pet!

eggs on their shoes into their homes. It is strange to think that though we take great care over personal hygiene we allow our pets to foul the environment.

This does *not* mean you shouldn't keep pets. But it does mean you should be aware of the responsibility of looking after them. Dogs and cats must be frequently 'wormed'. Small children must be taught not to kiss their pets on the face. All pets should be taken to the vet's surgery at regular intervals. Pet-owners in cities and towns need a small patch of ground where they can bury their pet's faeces. Dogs should be well trained, because in 1975 in Britain 150 000 people were bitten by dogs and at least 1500 road accidents were caused by dogs suddenly dashing onto the road. Small children are sometimes frightened of dogs and should be protected from them.

Safe disposal of waste

Sewage is the name used for *liquid waste*. Liquid waste includes urine and faeces from the lavatory, used water from sinks and baths and liquid waste from factories and industries.

Refuse or **garbage** are the names used for *solid waste*. Solid waste includes food scrapings and left-overs from the kitchen, papers, cans, boxes, newspapers, and all rubbish from offices, schools, factories and industries.

GETTING RID OF SEWAGE

A trap is a U-bend or S-bend in a pipe where fresh water lies. The fresh water acts as a seal, stopping any nasty smells from the sewer pipes coming back into the lavatory. Find out where else there are 'traps' in your home.

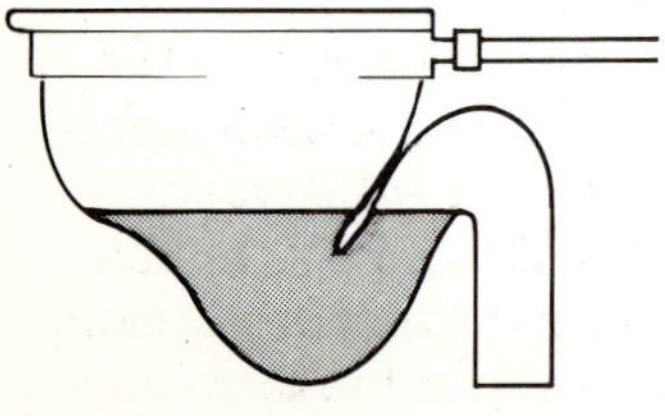

Fig 19.18 Diagram of a lavatory showing a water 'trap'

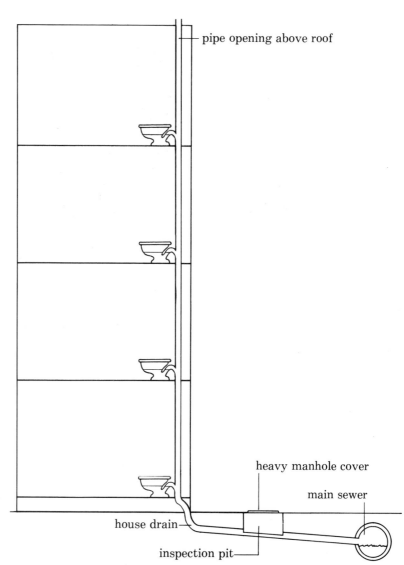

pipe opening above roof

heavy manhole cover

main sewer

Fig 19.19 The sewage system in a block of flats

house drain

inspection pit

Fig 19.20 Inside a large sewer

The sewage from buildings is carried away in pipes which meet under the ground at the main sewer. The inspection pit, used for cleaning blocked pipes, is always covered by a heavy, tightly-fitting manhole lid. Rainwater runs into the street gutters leading to the sewer pipes which spread like a network under the city streets.

TREATMENT OF SEWAGE

The raw sewage passes through a screen to remove large objects like glass or boxes which sometimes get into the sewers. This is dumped and buried. The raw sewage next goes into settling tanks where the heavier solid waste sinks to the bottom to form a **sludge**. The sludge is then pumped into huge vats where it is 'digested' which means broken down and made harmless by helpful bacteria and other micro-organisms (page 231). The treated sludge is dried and can be used as fertilizer or dumped out at sea.

257

Fig 19.21 An aerial photograph of a large sewage treatment works

The liquid raw sewage is run into huge ponds where it is stirred and aired by pumps. More helpful bacteria and other micro-organisms 'digest' the suspended matter in the liquid. The clean water is passed into filter beds where more bacteria and fungi purify it as it drains down through the filters. The **effluent**, the treated water, is then pumped into the rivers.

But two-thirds of the world's people do not have piped water. How do they get rid of their sewage? People living in lonely places simply go behind a bush and then bury their waste. Others use a chemical or earth closet – these are also used by campers and caravaners – and bury the contents of the pail or bucket. Untreated sewage like this must be buried away from shallow wells as there is a risk of infecting drinking water. The people who bury sewage must wash really well afterwards.

GETTING RID OF REFUSE

Household dustbins attract flies, wasps, cockroaches, ants and other pests. To prevent them becoming places for the spread of disease:

Line the inside of the dustbin, old newspapers will do.

Wrap up damp waste food, such as potato peelings, in paper.

Always keep the lid firmly on.

Put the dustbin as far away from the kitchen as possible.

Scrub the dustbin from time to time using a mild disinfectant.

The local authorities arrange for dustbins to be emptied at regular intervals. The lorries are specially built with lids to keep out the flies and a tip-up mechanism for dumping the rubbish. Refuse is either burned in huge **incinerators** or it is buried in large dumps and covered with earth.

People not living in towns can either burn or bury their refuse. If they bury it they must make sure the dump is dug very deep or rats will burrow down and use it for food.

Making sure the environment is a safe place to live is the responsibility of each person in the community. The public health authorities keep a constant watch and laws are passed to prevent us from polluting our towns and cities. Modern housing and good town planning to include parks and open spaces make a better and healthier environment for us all.

Write down all the ways in which the people living in picture b of Figure 19.22 are better protected against the spread of infectious diseases than are the people in picture a.

Fig 19.22 a Slum housing
b A Modern housing estate

a

b

Questions and things to do

Imagine you are taking a group of 9-year-olds camping for a week. *You are responsible for their health.* What sorts of foods would you take and why? Would you take milk? What would you do to make sure they only drank safe water? How would you protect them and the food from insects? What would have to be done to get rid of their sewage and refuse safely?

There is only space here for an outline of the ways in which a community keeps healthy by preventing the spread of infectious disease. Use the suggestions below to do a full health project on one of the topics. Before you begin, go to the town hall, the health education office or the local public health authorities to arrange visits and to collect as much information as you can.

Water

Draw a map of the world distribution of water. What are the main health problems of countries with little rainfall? Make a chart of last year's rainfall in your area. Find out the source of your town's water supply. Visit the water station to see how water is filtered and sterilized. Make a small model filtration bed and test it. Find out from the caretaker, or from plans of your home, where the water comes in from the 'mains'. Find out about water rates. Read the history of Dr John Snow (1813–1858) and the pump handle. What is meant by 'hard' and 'soft' water. Make a list of the advantages and disadvantages of each. Collect some rainwater and use it for washing. Find out more about water-borne diseases and their prevention.

Food

Draw a sketch or make a model of your local supermarket or food shop. On it, show shelves of food preserved in different ways. Write about 'display arrangements', 'shelf life', 'date stamping', 'perishable foods' and the temperatures in the frozen food cabinets. Visit a farm, a dairy, a milk bottling plant, a food factory or a food warehouse. Make notes on the care taken to prevent food becoming infected. 'Food poisoning happens more often in the summer months.' Use this sentence to write about the ways in which each member of the family can prevent this happening in their home. Read the history of Louis Pasteur.

Sewage and refuse

Visit a sewage works to learn more about how sewage is treated. Find out what method your local authority uses to get rid of its refuse. Draw a plan of your home or school showing where the waste pipes are. Why, after an explosion or earthquake, is there a very real risk of the people catching typhoid fever or cholera? Find out and describe how a cess-pit works. Write about the ways in which people can re-infect themselves from dirty health habits. Read the history of what Edwin Chadwick did in 1834.

House and town planning

You may prefer to do this project either on good housing or on town planning. If you choose housing you will need to find out about: choosing a healthy site; materials used for building, solid foundations, damp proof courses; different ways of heating, lighting and ventilation; safety factors; planning of kitchens for hygienic food preparation and storage; wash basins, baths, sinks and lavatories. Your house needs to be convenient for work, schools, hospitals, churches, shops and places of entertainment.

Town planning will include good public services such as gas, electricity, water, sewage and refuse disposal, transport, play spaces and recreational areas. You will need to find out about slum clearance, overcrowding, air pollution, pest control and other things which may cause the spread of disease.

Chapter Twenty

Infectious diseases 3: How we resist them

Why do some people catch colds more often than other people? And why do children get infectious diseases like mumps, measles and chicken-pox? How is it we recover quickly from some infections but that other infections make us seriously ill?

Our body has three main ways of resisting and fighting disease. We will call them the 'three lines of defence'. The first line is our skin, the second our white blood cells, and the third, the antibodies we make against particular infections. If germs do manage to get past our skin, then the second line of defence begins to attack them. And if the white blood cells aren't strong enough to destroy the germs, then antibodies are quickly made to help in the battle against disease.

Skin – the first line of defence

We know our skin acts as a barrier which stops micro-organisms from getting into our body (page 103). It is the most important line of defence because it is *always* better to prevent disease than to cure it. But there are two ways that micro-organisms can get past the skin barrier. They can get in through the natural openings and they can get in when we wound or damage the skin.

The natural openings have their own defence systems. Tears wash out the eyes, hair and mucus trap germs in the nose, saliva from the mouth, and hydrochloric acid from the stomach destroy germs in food. The reproductive and waste passages don't have such good defences so we must be extra careful over hygiene. Our ears are not listed in the diagram as we don't often get infections from outside, but they also have their defence system of wax and tiny hairs. (The lungs and breathing tubes have extra defences (page 55).)

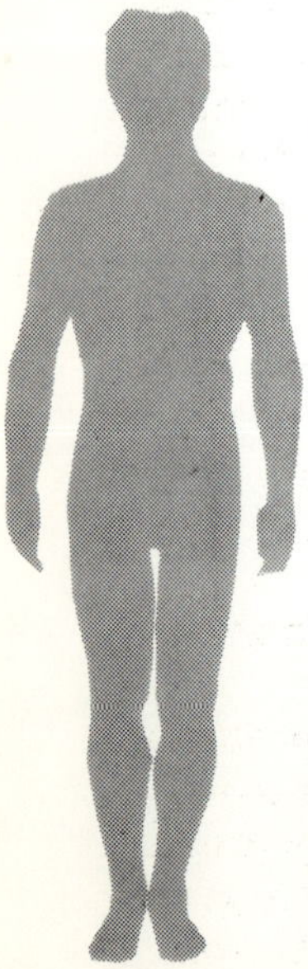

Natural openings

Eyes — contact
Nose — airborne
      foodborne
Mouth — waterborne
      contact

Penis
Vagina — contact
Anus

Breaks in the skin

Insect bites — pierce the skin
Cuts and scratches — open the skin
Fungus infections — land on the skin
Squeezing spots — opens the skin
Hookworm and scabies — burrow under the skin

Fig 20.1 Gaps in the skin defences

262

Fig 20.2 Treat all cuts properly

When we break open our skin a blood clot is made to plug up the opening (page 36). This takes a bit of time, and while it is happening micro-organisms are likely to get into the wound. We should wash all cuts under running water to clean out any dirt and germs, dab with a mild antiseptic to kill off any germs still in the wounds and cover with a clean plaster to stop more germs from getting in. Treating cuts properly also stops any pus which may be made from spreading to other people and infecting them.

We help our skin in its first line of defence by:

a. washing it regularly and keeping it clean;
b. making sure we have high standards of personal hygiene;
c. treating all cuts properly.

White blood cells – the second line of defence

Imagine your collar has scratched open the skin on the back of your neck. You feel the sore place with your finger and germs get into it. Turn to page 35 and study the way in which white blood cells act as a second line of defence. There are several kinds of white blood cells. Some poison the germs, some devour them and some clear up the mess of dead tissue and germs left at the end of the battle. If a lot of germs have got into the sore place and are breeding very fast, the bone marrow and lymph tissue speed up their production and enormous numbers of white blood cells are made and taken to the battle-ground.

Once the battle is won, the lymph capillaries take away the pus and debris to the nearest lymph node where it is filtered out and destroyed. The second line of defence is important for dealing with germs in the tissues and stopping them from getting into the bloodstream.

Antibodies – the third line of defence

When bacteria or viruses get into our body, they harm us in three ways.

1. Their 'bodies' are made of foreign (different) proteins from ours. The foreign proteins are poisonous to us.
2. As the pathogens feed, breed and excrete, they produce their own waste matter which is also poisonous to us.
3. They break down our living tissue, making our own cells harmful and destroying them.

The poisons from bacteria and viruses are called **toxins**. Some germs can be so toxic that the white blood cells are unable to destroy them. Antibodies, the third line of defence, have to be made specially to deal with them.

Antibodies are chemicals in our bloodstream. They deal with the toxins in many complicated ways by changing their structure and making them able to be devoured by the white blood cells. For each separate type of germ we have to make a separate and matching type of antibody. The antibodies we make against measles germs are no use against chicken-pox germs. When we recover from measles we are **immune**, or protected, from another attack (page 267). But we are still not immune to chicken-pox. And we will not be immune till we have caught the disease or been in contact with it and made antibodies against it. You will understand now why children get so many infectious diseases.

It takes a lot of extra energy to make more white blood cells and antibodies. Our general state of health is most important in the battle against infectious diseases. If we are not eating the right foods, are run-down, over-tired or taking other risks with our health, we don't have the reserves of energy we need. Our **resistance** to infection is low and this is one of the reasons why some people catch colds and other illnesses so often.

The course of an infectious disease

Even when you are fit, healthy and clean in your health habits and hygiene you may still get a serious infectious disease. Why is this? Some micro-organisms are so **virulent**, so full of powerful poisons, that they will cause illness no matter how careful we are. Once we are infected with a large number of virulent germs the battle against disease really begins.

THE INCUBATION STAGE

Often we don't know when virulent micro-organisms get into our body. We feel nothing unusual. It can be 2 days, 2 weeks or longer before we have any symptoms of disease. This is the incubation stage, the time it takes for the germs to invade the tissues, to multiply in vast numbers and to begin destroying our cells. Different diseases have different incubation times. We feel the symptoms of food poisoning within 12 hours of eating infected food. A cold takes between 2 and 5 days to appear and chicken-pox takes between 2 and 3 weeks. During the incubation stage we are **infective** (able to infect others) even though we may have no idea we have a disease.

THE FEVER STAGE

By this time the signs and symptoms of the illness can be seen and felt. The germs have got a strong hold on our

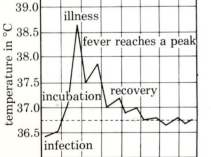

Fig 20.3 Temperature chart of a patient suffering from an infectious disease

tissues and are producing toxins and destroying our cells. Huge numbers of white blood cells and antibodies are being made and rushed to the site of the infection. All this extra work and the poisons from the germs cause the body temperature to rise. The heart beat and pulse rate increase, we get headaches and heavy sweating and, for a while, our minds may become quite confused as the battle rages inside us. During a fever we lose our appetite. All our energies are needed for the battle and so we don't use up any energy in digesting food. A person in a fever needs plenty to drink as so much water is lost in sweating.

THE RECOVERY STAGE
You can see from Figure 20.3 that our temperature drops as the germs are destroyed and we begin to recover. We are weak and exhausted, needing rest and sleep and a nourishing diet to build up our strength again. Our resistance to other infections is very low, so we should not go back to school or work till we have completely recovered. However, we do have plenty of antibodies against the disease we have just recovered from.

The course of a disease is infection, incubation, fever and recovery. We are able to infect other people at any time throughout these stages. Take extra care over your hygiene when you have been in contact with a person suffering from an infectious disease.

Immunity and immunization

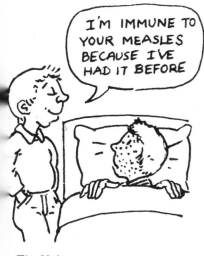

I'M IMMUNE TO YOUR MEASLES BECAUSE I'VE HAD IT BEFORE

Fig 20.4

WHAT IS IMMUNITY?
Immunity is being safe from the germs of any one particular disease because we have the matching antibodies to deal with them in our body. We get the antibodies by making them after we have been in contact with the disease. Chicken-pox and measles are nasty illnesses but not dangerous ones. Other diseases such as smallpox, whooping cough, diphtheria and tetanus are very dangerous. Hundreds of thousands of children used to die from them. The germs were so virulent that the children were not able to make enough antibodies to deal with them. Nothing could be done to save the lives of children who caught these killer diseases. Parents knew that not all of their children would live to become adult. And then came a fantastic breakthrough in medicine. Immunization was discovered.

EDWARD JENNER
An English country doctor, Edward Jenner (1749–1823), noticed that people who caught cowpox, a mild illness from cows, didn't catch the dangerous disease of smallpox. Having had cowpox seemed to make them immune to the germs of smallpox. To test this he took some pus from a

265

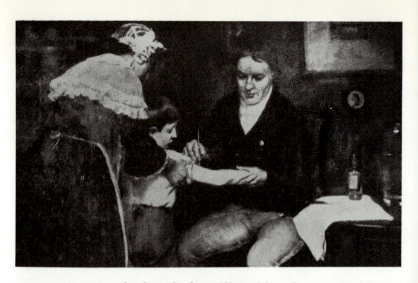

Fig 20.5 Dr. Jenner injecting James Phipp's 1796

cowpox sore on the hand of a milkmaid and scratched it lightly into the arm of an 8-year-old boy. The boy, James Phipps, got cowpox but soon recovered.

Then, probably feeling very anxious, Jenner injected some pus from a smallpox sore into the boy's arm. To everyone's great relief the boy didn't get smallpox. Dr Jenner had proved we could trick our defence system into making antibodies against a serious disease by injecting the weaker germs of the same kind of disease. James Phipps was immune to smallpox.

It was called **vaccination** because 'vacca' is the Latin word for 'cow'. At the time, a lot of people were horrified by the idea of actually putting a disease into a healthy person. In spite of the fuss, so many people were vaccinated against smallpox that the disease in Britain was almost wiped out. In 1898 it was made compulsory for every person to have the smallpox vaccination. This was stopped in 1948 as it was no longer necessary. Smallpox, which brought such terrible suffering to mankind, is being wiped out all over the world because of Dr Jenner's brilliant idea.

WHAT IS IMMUNIZATION?

Doctors and scientists began to work at finding different immunizations against other killer diseases. This is a very difficult thing to do, as there is a slight risk of giving a healthy person a disease. The germs must either be made very, very weak or they must be killed before they can be used. In some diseases, only the toxins from the germs are safe to use. They are made into a vaccine which is then injected into us. We don't get ill. Our body defence system begins to make antibodies against them. Later, we get another injection and we make more antibodies. We still don't get ill. Soon we have enough antibodies to make us immune if we come into contact with the disease itself.

Immunization is a very clever way of getting people to make antibodies against diseases – safely. This is done by 'tricking' our defence system into making antibodies without ever having to catch the disease first.

HOW IMMUNIZATION CAN WIPE OUT A DISEASE

Diphtheria was a killer disease in Britain until a vaccine was made against it. In 1935 a campaign was started to immunize all children with the vaccine. These are the results:

Year	Number of cases	Number of deaths
1935	59,000	2,875
1942	41,404	1,827
1946	18,294	444
1954	182	9
1970	22	–

Can you find out the records for the disease poliomyelitis?

ACTIVE IMMUNITY

Vaccine can be injected into us, or vaccinated into us, or taken by mouth. Our defence system gets to work to produce the matching antibodies. As such tiny amounts are given at a time, it takes quite a while before we have built up enough antibodies. We are not immune to a particular disease until we have had *all* the 'jabs', 'shots' or sugar lumps.

This kind of immunity is active immunity because we have *actively made the antibodies ourselves*. Once the defence system has learned how to make a particular antibody, the information is stored – though we don't yet know how this is done. Antibodies don't last for ever, they wear out, but we are still immune once we have the stored information on how to make more antibodies.

PASSIVE IMMUNITY

This kind of immunity is used for emergencies. The ready-made antibodies are taken from a person or animal who has recovered from the disease and made into a vaccine. It is called passive immunity because *we don't make the antibodies ourselves*, we get them already made. Tetanus is a killer disease which we may get from a deep wound or deep puncture in our skin. Immediate help is needed so an injection of ready-made antibodies is given. They can get to work at once to deal with the tetanus bacteria. Passive immunity is also used when there is an outbreak of a disease. Ready-made antibodies only last from 4 to 6 weeks. After this time we have no more protection against the disease. As we have not made the antibodies ourselves, there is no store of information our defence system can use. We need to be injected with the weakened germs or

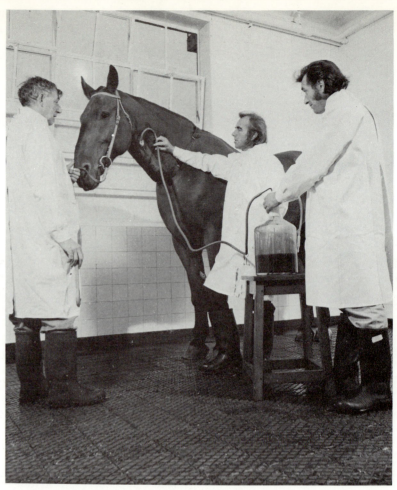

Fig 20.6 Blood being taken from a horse so that the antibodies in it can be used for passive immunization

their toxins so we can make our own active immunity.

A baby is born with passive immunity he gets from his mother. For the first months of his life he is kept safe by these antibodies. If he is breast-fed, he goes on being protected as antibodies are passed in human milk. When he is weaned, the antibodies wear away and he is left without any immunity of his own. He must be given his first active immunizations. These will be done at a Child Health Centre, a clinic or the doctor's surgery. Study Figure 20.7 now.

Why is it so important that all girls have the rubella injection at this age? Why is the smallpox vaccine not given any more? Imagine you have a friend who can't be bothered to take her baby for his Dose III injection. Write down how you would explain why it is necessary that her baby has *all* his injections.

Different immunities last for different lengths of time. A 'booster' dose is given when there is a risk that the store of information might be getting weakened.

As yet, we haven't found out how to immunize against all infectious diseases. And not all diseases give us

Age	Vaccine
5 to 6 months	Diphtheria, whooping cough and tetanus given in one injection. Polio taken by mouth. Dose I.
6 to 8 weeks later	Diphtheria, whooping cough and tetanus given in one injection. Polio taken by mouth. Dose II
4 to 6 months later	Diphtheria, whooping cough and tetanus given in one injection. Polio taken by mouth. Dose III
About 3 years old	Measles vaccine
About 5 years old	Diphtheria and tetanus 'booster' injection. Polio booster.
Between 10 and 13	B,C.G against tuberculosis for children who are not immune. Special tests show if there is an immunity.
Girls 10 to 13	Rubella vaccine against German measles.

Fig 20.7 Children's immunizations

immunity after we have recovered from them. How many colds do you catch in one winter! The serious STD's, syphilis and gonorrhoea, don't give us immunity from another attack. Vast amounts of time, talent and money are spent on research into disease and the body's defence system. New immunizations are likely to be discovered at any time.

IMMUNIZATION FOR TRAVELLERS

Certain diseases are **endemic**, they happen regularly, in certain parts of the world. Cholera is endemic in India, bronchitis is endemic in Britain. Other diseases, like 'flu or measles, are fairly common everywhere. (Re-read page 57 to make sure you don't get muddled between endemic and epidemic.) If we go to a country with different endemic diseases we have no immunity against them and will get

RULES

The vaccine used shall meet the requirements laid down by the World Health Organisation.

The validity of this certificate shall extend for a period of six months, beginning six days after one injection of the vaccine or, in the event of a revaccination within such period of six months, on the date of that revaccination (see Note 1 below).

The approved stamp mentioned overleaf must be in a form prescribed by the health administration of the territory in which the vaccination is performed (see Note 2 below).

This certificate must be signed by a medical practitioner in his own hand; his official stamp is not an accepted substitute for the signature.

Any amendment of this certificate, or erasure, or failure to complete any part of it, may render it invalid.

Le vaccin utilisé doit satisfaire aux normes formulées par l'Organisation Mondiale de la Santé.

La validité de ce certificat couvre une période de six mois commençant six jours après une injection de vaccin ou, dans le cas d'une revaccination au cours de cette période de six mois, le jour de cette revaccination.

Le cachet autorisé doit être conforme au modèle prescrit par l'administration sanitaire du territoire où la vaccination est effectuée.

Ce certificat doit être signé par un médecin de sa propre main, son cachet officiel ne pouvant être considéré comme tenant lieu de signature.

Toute correction ou rature sur le certificat ou l'omission d'une quelconque des mentions qu'il comporte peut affecter sa validité.

NOTES

1. Method of dating.

Misunderstandings have arisen as to the date of issue, and therefore the period of validity, of International Certificates of Vaccination, due to differences in national or other practice of recording dates: for example, the 10th August, 1952, may be written as 10 Aug., 1952, or Aug. 10, 1952, or 10.8.52, or 8.10.1952.

These misunderstandings can be avoided if dates on International Certificates are always written thus:—

the **day** should be placed first in **Arabic** numerals;
the **month** should appear second in **letters**;
the **year** should come last in **Arabic** numerals.

The above example would then appear as "10 August, 1952."

2. Approved Stamp.

Approved stamps have been prescribed for the United Kingdom. If the vaccinator is not himself an authorised user of a stamp, the person vaccinated must take or send the certificate for stamping to a Local Authority.

In England and Wales, this is the **Town Council, Urban District Council or Rural District Council,** and in Scotland the Council of the County or large Burgh, in whose area the vaccinator practises.

In Northern Ireland this is the office of the Health Committee for the County or County Borough in whose area the vaccinator practises.

FOLD HERE

Form of certificate prescribed by the International Health Regulations and supplied, for completion in the United Kingdom only, by the
DEPARTMENT OF HEALTH AND SOCIAL SECURITY
Scottish Home and Health Department, Edinburgh
Ministry of Health and Social Services, Belfast
Welsh Office, Cardiff

INTERNATIONAL CERTIFICATE OF VACCINATION OR REVACCINATION AGAINST CHOLERA

CERTIFICAT INTERNATIONAL DE VACCINATION OU DE REVACCINATION CONTRE LE CHOLERA

This is to certify that
Je soussigné(e) certifie que

name / *nom* SUSAN ANN MORAN

date of birth / *né(e) le* 30 MAY 1955 sex / *sexe* FEMALE

whose signature follows
dont la signature suit

S.A. Moran

has on the date indicated overleaf been vaccinated or revaccinated against cholera.

a été vacciné(e) ou revacciné(e) contre le choléra à la date indiquée au verso.

IMPORTANT: See the Rules and Notes on the back.

This certificate, if folded along the central vertical line, can be kept in a passport (with a rubber band).

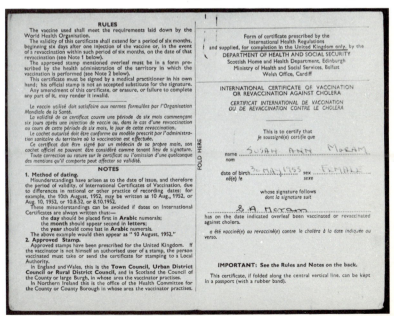

Fig 20.8 An international certificate of vaccination

very ill indeed. The four diseases we must be immunized against are the typhoid fevers, smallpox, yellow fever and cholera. You must be able to prove you have been immunized by showing an up-to-date certificate of vaccination. Without the necessary certificates you could be stopped from entering a foreign country or from re-entering Britain until you had been vaccinated.

Notifiable diseases

Serious infectious diseases must be reported to the **Community Physician** at once. They are called **notifiable** and some of them are:

Cholera	Smallpox
Diphtheria	Tetanus
Food poisoning	Tuberculosis
Malaria	Typhoid
Measles	Whooping cough
Poliomyelitis	Yellow fever

Some infectious diseases which are not notifiable are: common cold, influenza, pneumonia, chicken-pox, rubella and mumps. Gonorrhoea and syphilis, though very serious and highly infectious diseases, are not notifiable in Britain. They are notifiable in many other countries. The most recent disease to be made notifiable is Lassa fever.

The Community Physician is a doctor with special training in the health and hygiene of the community. When a notifiable disease is reported, part of his or her work is to trace the source of the infection. It could be an infected water pipe which is polluting the drinking water; a careless shop assistant infecting the food; a person with infected faeces using a public convenience without proper standards of hygiene. It could be so many things that a very careful history is taken about the patient. Where has

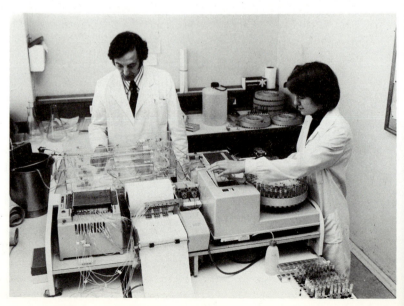

Fig 20.9 Samples of blood being analysed in a laboratory

he been? Whom has he been in close contact with? What has he eaten, drunk? And so on. Samples of his blood, urine, faeces or nose and throat swabs are taken and sent to a laboratory to be **analysed**. Trained technicians test them to find out what micro-organisms are causing the disease. Why do you think the standards of hygiene in a laboratory are so high?

Treatment by drugs (drug therapy)

Another line of defence in the battle against disease is the use of drugs. Drugs are used as a cure for illness. We take drugs when we have a disease.

The most dramatic breakthrough in drug treatment happened in 1929. Alexander Fleming, a British professor, noticed that some fungus had landed on a culture of bacteria he was studying. He also noticed that the bacteria had disappeared from the area around the fungus. He wondered if there was some 'thing' from the fungus which was destroying the bacteria. He grew the fungus and tested it with different bacteria. He was right. The fungus had the power to destroy many kinds of bacteria which caused human disease. He called it **penicillin**.

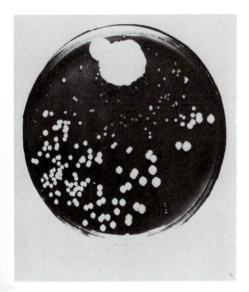

Fig 20.10 The Petri dish in which Fleming discovered penicillin

Fig 20.11 Sir Alexander Fleming in his laboratory

But it wasn't till 1940 that penicillin was used as a drug. It was given to a policeman suffering from blood poisoning in an Oxford hospital. At first the policeman began to recover, but there was not enough of the drug to give him. When the supply ran out he became very ill again and died soon after. It was decided that vast amounts of penicillin should be made so a laboratory was set up in the USA as Britain was at war. Very soon penicillin was being called 'the miracle drug'. It was used all over the world and it is not possible to work out how many lives have been saved, or how much sickness and pain has been cured by its use.

Later on, it was found that a few people are **allergic** to it, that some harmful micro-organisms get a resistance to it and that it is of little or no use against viruses. There are not yet drugs against diseases caused by viruses. After the fantastic success of penicillin, scientists all over the world began to make new discoveries on other strains of fungus. We now have many powerful drugs, as well as penicillin, which are made from helpful micro-organisms. They are called **antibiotics** and the most recent ones are chemically treated as well.

Drugs are used as a cure for disease: we do not usually need them for minor ailments. But many people believe they should be given a bottle of medicine or some tablets every time they visit their doctor. They feel cheated – as if they hadn't been looked after properly – if they are not given drugs. It doesn't make sense to feel this way. Most doctors are over-worked and they want their patients to get better as quickly as possible. Of course they will give drugs if they are needed. When you go to your doctor, don't feel cheated if you are not given drugs. Listen carefully to the advice and instructions on what you should do to get better. When you get home, do exactly what you have been told. Your body can cure itself of minor ailments if you treat it properly.

Fig 20.12

Disinfectants and antiseptics

These are the names given to strong **disinfectant** and mild **antiseptic** solutions which have the power of killing germs. Disinfectants are used in sinks, drains, lavatories, and so on. Antiseptics are used to clean and swab human tissue. They are another essential weapon in our fight against disease.

But there was a time when little was known about them. Only a hundred years ago hospitals were filthy places. They were overcrowded; they smelled of vomit, pus and decay; rats, mice and cockroaches lived in the soiled bedclothes and the refuse from wounds and dirty bandages. A healthy mother going to hospital to have her baby might die, not from the birth, but from the infection she would catch from the midwives, the doctors and her surroundings. People dreaded going to hospital because of the high risk of catching a deadly infectious disease. Joseph Lister (1822–1912), working as a surgeon in a Scottish hospital, was shocked at the numbers of people who died after an operation. The surgeons used to keep their old coats to wear during the operation and these coats became soaked in dried blood, discharge and pus from the wounds. The equipment in the operating theatre was wiped down with an old cloth.

Lister noticed that a wound healed quickly if there was no pus in it. He began to think there might be a link between pus and infection. After reading Pasteur's work, he started a campaign against dirt. Solutions of carbolic acid, a very powerful disinfectant, were sprayed everywhere and the surgeons' clothes, instruments and hands were washed in it. The results were wonderful. The death rate dropped as wounds healed without pus and the patients recovered.

Nowadays, hospital operating theatres are made as nearly germ-free as possible. All the equipment and the clothes are sterilized. This is done by steaming or boiling them at very high temperatures. Even the patient's skin, where the cut is to be made, is swabbed with an antiseptic first. It's not possible to boil up the doctors and nurses! But, as you can see from this photograph, and from Figure 18.3 (page 233), they make themselves as germ-free as possible.

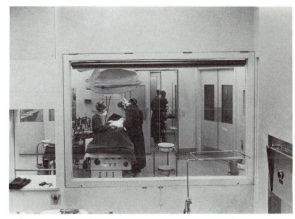

Fig 20.13 An operation in 1882 *after* Lister's improvements. Notice the carbolic spray and ordinary clothes
Fig 20.14 A modern operating theatre

Many things have been discovered to help us win the battle against harmful micro-organisms, pathogens. We are no longer helpless against infectious diseases. We can now hope to fight them in many different ways:

by educating people about the importance of personal hygiene;

by preventing the spread of disease;
by a high standard of public health services;
by immunization and vaccination;
by the new drugs;
by sterilization and the use of disinfectants and
antiseptics;
by health education in schools.

**Questions and things
to do**

As there are so many new words, it will be a useful memory
exercise if you write each of the following in a separate
sentence to show its meaning: pathogen, micro-
organism, virulent, incubation, toxin, immunization,
quarantine, vaccine, endemic, antibody, chemotherapy,
antibiotic, disinfectant.

1. What are the ways in which micro-organisms can get into
   our body?
2. What can we do to prevent this happening?
3. In your own words, describe what happens during the three
   stages of an infectious disease.
4. What are the three ways pathogens harm us once inside
   our bodies?
5. When we have a fever why (*a*) does our temperature rise
   and (*b*) do we lose our appetite?
6. Explain clearly the differences between active and passive
   immunization.
7. Copy out and learn the vaccine chart for children in
   Britain.
8. Why is it so important we have the full course of injections
   and the booster injections?
9. Choose any four notifiable diseases and find out their
   incubation periods.
10. What is pus?
11. Measles, mumps, chicken-pox and German measles are
    childhood diseases. How many people in your class have
    had them? Find out and write about what types of pathogen
    causes them, how they are spread, their incubation period,
    their signs and symptoms. How would you nurse a patient
    through one of these four diseases?
12. Write a few lines on the work of each of the following:
    Jenner, Pasteur, Fleming, Lister, Simpson, Salk.

Chapter Twenty-one

Health and responsibility

Each country has the responsibility of looking after the health of its people.

World health

The more money a country has, the more it can spend on health care, health services and health education. The people get stronger and healthier. They can work hard because they are free from illness. They make more money for their country so there is more money to spend on health.

The poorer a country is, the less it can spend on looking after the health of its people. The people may be weak from malnutrition, their body defences may be run down from many attacks of disease. They do not have enough energy to work hard as they suffer so much illness. They cannot make enough money for their country so there is even less money to spend on health.

OVERSEAS AID

Poorer countries need help. Some of the wealthier countries, particularly the USA, give a great deal of money in **overseas aid**. More important, they send out doctors, nurses and health educators. They send out people who can introduce modern farming methods, who can improve the water supplies and sanitation, who can teach mothers how to protect their families from disease, and so on. The best way to give aid to the poorer countries is to help them to help themselves.

THE WORLD HEALTH ORGANIZATION (WHO)

This was set up to look after all world health problems and to help control the spread of disease. Its headquarters are in Geneva in Switzerland and there are offices in India, Africa, Egypt, Denmark, the USA and the Philippines. Some of the work done is listed below.

Information to all countries about outbreaks of disease

Information about the risks from radiation

Information to all countries about new discoveries in medicine

Information about new vaccines and tests on them

Information on food and water studies

Protecting maternal and child health

Information on human genetics

Improving water supplies and sanitation

Information on healthy environments

Setting up international quarantine rules

Research into mental health

Research into new drugs and drug addiction

Research into the causes of accidents

Research into health problems of industrialised countries

Help to underdeveloped countries

Training people to teach health education

Campaigns to control malaria, smallpox and other serious infectious diseases

Research into addiction and drug control

Community health

The Department of Health and Social Security (DHSS) is responsible for the health and welfare of the people in Britain. It plans and organizes the **National Health Service**, the **Social Services** and the **Social Security Benefits.**

While working through this book, you have been asked to visit some of the public health centres to find out more about the work done to help the community. If you haven't had time to do so during the school term, use part of your holidays for a special study or project on at least two different types of community care. (As one of your studies you may like to choose to find out more about the work of the DHSS itself.)

There are also many **charities**, privately run organizations, which help in the care of the sick, the elderly, the handicapped, the homeless, invalid children, and so on. They also raise money to help overseas charities. They are always in need of extra people to help out. By doing voluntary unpaid social work, by helping people when they are in trouble or in need, you may find a future career for yourself which will bring you much pleasure and satisfaction.

SOME WAYS OF COMMUNITY CARE

Ante-natal clinics

Genetic Counselling Service

Maternity hospitals

Child Welfare Clinics

Home nursing

Vaccination and immunization

School Health Services

Drug control

Care of the elderly

Care of the mentally handicapped

Care of the physically handicapped

Care of the sick at home

Food inspection

Care of children in need
Prevention and control of infection
General Practitioners
Hospitals
Doctors, nurses and technicians
Dentists, opticians and chemists
Ambulance Service
Family Planning Clinics
Special STD clinics
Chiropodists
Health education

Safe water supplies
Sewage and refuse disposal
Town and country planning
Housing
Pest control
Noise control
Control of air pollution
Safety Officers
Blood Transfusion Centres
Mass X-ray units
Anti-smoking Clinics
Slimming clubs
Alcoholics Anonymous
Mental health

Personal health

We can make fantastic discoveries in medicine; we can prevent the spread of disease; we can make living conditions clean and safe; we can do many, many things to improve the health and welfare of people. But we cannot stop people from taking risks with their health. They are free to chose whether to use or abuse their mental and physical health.

You have this freedom to choose. Small children need to be helped to learn how to look after themselves. But from adolescence onwards, you are the one person who is responsible for your health.

'I am the master of my fate:
I am the captain of my soul.'

This book may have given you some understanding of the incredibly 'clever' workings of the body and mind. How marvellously the systems all function smoothly together to produce that most fascinating of all living things – a human being.
Hamlet said:

'What a piece of work is man! How noble in reason! How infinite in faculty! In form and moving how express and admirable! In action how like an angel! In apprehension how like a god! The beauty of the world! The paragon of animals!'

We don't often think of ourselves in quite such lofty and splendid terms. But we are quite marvellous. We have the gift of life.

Index